M000117962

THE BIG

COMMITMENT

THE **BIG**
C O M M I T M E N T

SOLVING THE MYSTERIES OF YOUR

ERP IMPLEMENTATION

JOEL PATTERSON

ForbesBooks

Published by ForbesBooks, Charleston, South Carolina.
Member of Advantage Media Group.

ForbesBooks is a registered trademark, and the ForbesBooks colophon is a trademark of Forbes Media, LLC.

Printed in the United States of America.

10 9 8 7 6 5 4 3 2 1

ISBN: 978-1-94663-352-1
LCCN: 2020909612

Cover design by George Stevens.
Layout design by Wesley Strickland.

This custom publication is intended to provide accurate information and the opinions of the author in regard to the subject matter covered. It is sold with the understanding that the publisher, Advantage|ForbesBooks, is not engaged in rendering legal, financial, or professional services of any kind. If legal advice or other expert assistance is required, the reader is advised to seek the services of a competent professional.

Advantage Media Group is proud to be a part of the Tree Neutral® program. Tree Neutral offsets the number of trees consumed in the production and printing of this book by taking proactive steps such as planting trees in direct proportion to the number of trees used to print books. To learn more about Tree Neutral, please visit **www.treeneutral.com**.

Since 1917, Forbes has remained steadfast in its mission to serve as the defining voice of entrepreneurial capitalism. ForbesBooks, launched in 2016 through a partnership with Advantage Media Group, furthers that aim by helping business and thought leaders bring their stories, passion, and knowledge to the forefront in custom books. Opinions expressed by ForbesBooks authors are their own. To be considered for publication, please visit **www.forbesbooks.com**.

To everyone brave enough to embark upon an
ERP implementation and my mom.

CONTENTS

ACKNOWLEDGMENTS

THERE ARE SO MANY people who have influenced and contributed to the stories and feelings behind this book. Rather than try to identify a select few, I realize I've stolen something from every person at The Vested Group (and beyond!) and each deserves recognition:

Alexa Nguyen for demonstrating that there is always time to experience more.

Allison Davis for reminding me that there's no end to the strength that comes from unique and diverse backgrounds.

Aprille Tenorio for your rock steady approach to everything I throw at you, making me laugh more than anyone has a right to, putting your unmistakable mark on this book, contorting words into backbends in ways never intended, and most of all, being my friend longer than I deserve.

Austin Pretsch for showing me that there's no substitute for genuine happiness.

Brandon Fuller for proving there's always someone *way* smarter than me in the room.

Brittany King for showing that you don't have to be old in years to be wise in thought.

Bryn Harvey for not holding first impressions against me.

Casey Andrews for confirming that believing in the person is more important than the resume.

CJ Carter for reminding me that there's always someone out there that is destined to do my job better than I can.

Cedric Carter for being the oh-so-important counterbalance.

Chris Johnson for providing more content than I could ever retell on these pages. Oh, and Kelly Clarkson.

Chris Mixon for exuding the quiet confidence reserved strictly for legitimately brilliant minds. Plus dad jokes.

Cindy Cline for showing that some of us will always be hippies at heart.

Consuelo Gonzalez for your tireless spirit and commitment to excellence.

Daniel Keough for being there from Peoria to Plano and all the stops in between—please know that you and your family will always be part of mine.

Daniel Webb for graduating from the megaphone and finding your own voice.

David Robertson for confirming that people with large heads have big brains and yours leads the pack.

Debbie Pollard for teaching me that a charming Texas drawl can get shit done.

Denise Scarpa for mentoring me through a variety of topics including fitness, nutrition, and the importance of high-quality stilettos.

Dimple Patel for your fearlessness when you took a chance on us.

Dominick Stives-Mitchell for bringing your gentle approach to people and commitment to active listening.

Fergus Reynolds for being the happiest man I've ever known in addition to making every encounter an opportunity to learn.

Gabe Nill for leading us through the financial wilderness on two wholly unique safaris.

Hannah Wulz for never being afraid to be you even when you're right in the middle of the hiring process.

Heidi Convery for strengthening our strengths and inspiring us to learn more about each other.

Hillary Patterson for never letting me get away with anything— well, almost anything.

JP Mack for reminding me that you don't have to say a lot to have an influential voice.

Janelle Rodriguez for proving that you are always successful when you relentlessly pursue your goals.

Jason Lavender for bringing your A-game right out of the chute at work and at play.

Javeria Ahmed for demonstrating that accountants are indeed capable of bringing the sass with a smile every day.

Javier Magana for showing that a measure of wit and quiet charm allows you to get away with anything.

Jayne Gardner for shining a light on the path of my authentic self and kicking my ass when I needed to move forward.

Jennifer Polk for your never-ending support and empathy, putting up with my shit daily, and owning the fact that technology hates you.

John Keel for communicating with a vocabulary that consistently sends me to the dictionary.

John Mack for teaching me ways to lead with compassion—your people and clients love you.

Jon Brower for the memories that will never be forgotten and a sales mind that never forgets anything.

Jonny Zielinski for proving that it's possible to smoothly alternate between comic relief and sage advice in the same conversation.

Joseph Lang for continuing to prove that it's always worth talking people out of an offer letter when you know the recruit is a bad ass.

Josh Barnett for consistently wowing me with your vast knowledge of music, random trivia, and Dallas facts all while endearing yourself to every client you meet.

Karl Martinsson for showing us new ways a technical developer can surprise through baking, chili contests, and Grand Master-level Lego building.

Katherine Staiger for showing me that an unassuming, sweet smile can be a critical tool in rallying a team and making things happen.

Kimberley Schwab for providing a unique and entertaining perspective on almost every scenario imaginable.

Lucia Capurro for keeping that Uruguay crew in line and consistently delivering a quality product.

Luke Stebbins for showing that the old school methods of initiative and hard work are timeless.

Malcolm Williams for bringing a cool swagger and quiet intelligence to the office every day.

Maria Magana for reminding me to be brave and to believe in myself when identifying future goals.

Martin Berguer for continuing to show that exceeding expectations and making clients happy is possible from 5,300 miles away.

Matt Laban for always demonstrating patience and breaking down complex ideas to those of us with simpler minds.

Matthew Gregory for showing me that being proactive and sharing ideas benefits everyone around you.

Matthew Luna for surprising me and showing that it is indeed possible to smile even when you're the brunt of Hillary Hazing.

Mike Pope for being a mentor and leader in ways you may not even realize including a quiet confidence that oozes experience and trust.

Muzzamil Asif for demonstrating the consistency and temperament that will make you a star regardless of the situation.

Neeraj Sharma for your unwavering commitment to excellence, always doing the right thing, and defending your title as Nicest Man Alive.

Pablo Dominguez for demonstrating the patience of a saint waiting for us to catch up with you.

Patrick King for proving once again that interns and baseball players develop into the best consultants.

Rian Johnson for proving once again that first impressions are not always accurate especially when you have the flair of brilliance.

Ross Cortese for inspiring us to take a hard look at Cowboy Consulting as a new service line.

Shaunna Black for providing a valuable nugget every time we speak.

Shelley Fopiano for fearlessly attacking every challenge put in front of you and never once filtering your feedback.

Steven Smith for not overtly making fun of my code that you are probably still fixing.

Stevie Patterson for your love and support even when you're putting up with a demanding old man.

Terry Cole for always demonstrating your passion even when it's not the simplest path.

CORE VALUES

- Always Do the Right Thing

- Give Them Your Shirt

- All In

- Own It

- Enjoy the Ride

- Appreciate That Failure Results in Learning

AUTHOR'S NOTE

I'VE BEEN INVOLVED in business and systems consulting for more than two decades, and over that time I've had the opportunity to work for some of the largest and best-known consulting firms in the world. I was first introduced to the enterprise resource planning (ERP) industry in 1998, implementing Oracle E-Business Suite for a client in Ponca City, Oklahoma. As you might imagine, the technology and expectations were a little different then. We focused a lot on software configuration and making sure it worked properly. As things have matured, the projects I've worked on have shifted to more of a focus on business process and people development. The system is there to be a vehicle for change for an organization, and being part of that change is what makes our job fun.

While I am proud of my pedigree and track record, I am prouder still of my current position as the founder of The Vested Group, an award-winning consulting group specializing in the implementation and support of a world-class, cloud-based business management software package called NetSuite. Founded in 1998, NetSuite as of 2020 is used by over eighteen thousand organizations worldwide. Beyond leading NetSuite ERP implementations, we offer our clients strategic advisory services, software implementation expertise, multi-channel ecommerce solutions, and customized development projects.

In other words, we help put the system, processes, and training in place to provide business leaders with reliable and actionable information.

As founder of a company that helps clients better and more accurately understand their own business, I obviously recognize the importance of providing our current and future clients with an authentic view of our core values, our business philosophies, and our vision for the future.

At the same time, I am equally aware of the power of storytelling. I know a story can best show the nature of the services that our firm offers, describe the relationships we enjoy with our clients, and highlight what makes us different and unique, including our all-star team of consultants, whom I believe collectively distinguish The Vested Group from all other firms in the field.

We have helped enough companies solve the mysteries of ERP implementation to know that it is tough work, but there is a successful path out there if you are willing to make "The Big Commitment" to do what it takes. To help explain this commitment and what it takes, I decided to write a book.

When I was in college, I read *The Goal*, by Eliyahu M. Goldratt. I'll never forget marveling at how he was able to take a complex topic and make it simple by telling a story. That is my hope for this book. It's not an attempt to wow you with consultancy speak or stories of greatness but to simply show you that something as complex as an ERP implementation can be successful by following some simple steps. Just to be clear, simple does not equal easy. You must be willing to do the work, have the difficult conversations, and lead your team through a challenging period they will not likely want to repeat anytime soon. That's why we refer to it as The Big Commitment.

Like any founder of a company, I wear many hats at The Vested Group, but I remain directly and actively involved in the day-to-day management of a company that has enjoyed dramatic growth across the board. In that regard, I'm just like anyone else with a business in that I want to tackle the challenges in front of us and outpace the growth that we have planned. One of the main ways we achieve this is by leveraging NetSuite in every possible area of our business, which reinforces the fact that it is important to our clients that we are "users" like them.

But while wearing this founder hat, I've come to realize that business, as a whole and across the full spectrum of industries, has changed a great deal in the past twenty years or so. A *great* deal.

It's not simply that emerging technologies have transformed every market into a much faster track than the ones our predecessors had to run on or that corresponding developments have seen every industry become more competitive—although those observations are certainly accurate.

And it's not just that these amazing advances in technology have made the business world exponentially more complicated and have necessitated that every forward-savvy enterprise develop a long-term relationship with a digital business expert to ensure they are taking the fullest advantage of the incredible opportunities that exist—although all that's absolutely true, too.

No, the simplest of corporate tasks and day-to-day business processes have become significantly more involved, too. In today's business environment, all C-level executives face intensifying pressures and rising responsibilities, including an obligation to read all the time just to stay on pace with everything that's happening in the world.

Emails.

White papers.

Industry reports.

There's so much reading material to sort through in a single day at the office that just staying current seems like it has become a full-time job in itself.

Believe me, I understand.

So, while I'm eager to tell you all about The Vested Group and all of the unique services and products we provide to our clients to empower and enable them, the last thing I want to do is add to that ever-growing mountain of reading material that may or may not have true value.

That was our dilemma: every one of us here at The Vested Group wants these projects to be smooth and not unduly burdensome for our clients—at least, not any more inconvenient than it absolutely has to be—and so, we wondered how to get people to learn about some of the things we're doing for our clients *without* making that endeavor just another chore to be checked off on the day's endless to-do list (no one likes those).

It was a puzzler.

But if we at The Vested Group pride ourselves on anything, it's our unbridled creativity and our ability to come up with the perfect (even if unexpected or nontraditional) solution to absolutely every challenge our clients may be facing in their businesses. We like to refer to this as *TVGenius,* a fun term that describes who we are, what we do, and why—as you'll first see in chapter 1.

So, relying on that inspired sense of creativity, we thought long and hard about the most effective way to tell our story.

We wanted to offer a real insight into our company and our core values; a snapshot, if you will, of the people who make The Vested Group the industry leader that it is and how that translates into a successful project for your organization.

We wanted to provide a layman-friendly explanation of ERP systems so that any reader could understand how utilizing these systems can offer them an integrated and upgraded view of their core business processes and enable them to collect, store, manage, and analyze data from these many business activities in a way that will position them to maximize overall performance across a wide array of metrics.

We knew we had an important message to get out there, and at the same time, we wanted to make its delivery something more than expected. We wanted to come up with a complete final product that carried our message and conveyed the necessary information, but in a way that was still entertaining and that busy people like you wouldn't find a chore but rather more of a treat.

Soon enough the answer became obvious to us: write a novel.

Not just another traditional, by-the-numbers business book, mind you—that's not The Vested Group style.

A novel!

Better yet, a mystery!

Because who doesn't love a mystery?

No, wait.

Even better, a comedic mystery.

Because who doesn't love a mystery *and* a laugh or two?

Right?

Maybe with an added element of human interest, too? With a happy ending?

A literary gumbo with our message as the roux bringing it all together.

And you must love that totally original concept, because here you are holding this book and reading along.

Which brings me back to my point … you're actually holding our book and reading something that you might otherwise have overlooked. And that's all that we can ask.

So, thank you.

No, thanks, really.

Because we understand how valuable your time is, and we appreciate your spending it to turn a couple of pages of the story that we have to tell.

The following is just that: a lighthearted mystery that is meant to entertain you, all while attempting to explain in simple terms how any organization can successfully implement a new ERP. We at The Vested Group have been there and done that for a wide array of clients, and our incredible team of professionals can do the same to improve your daily business activities, your customer interactions, and ultimately your bottom line.

I hope you enjoy.

And that you're informed and intrigued.

And that you come away with a number of important takeaways.

And, ultimately, I hope that you reach out to our team directly so that we can sit down face to face and discuss further all of the great things The Vested Group can do for you and your business.

—JOEL PATTERSON, JUNE 2020

PART ONE

THE

MYSTERY

CHAPTER 1:

THE MYSTERY BEGINS

THERE WAS NO SHORTAGE of security in place at the Carter Supply compound. A ten-foot chain-link fence encircled the entire perimeter of the hundred-acre property.

There were only two gaps in that fencing.

The first was a single entrance/exit along the eastern side for truck traffic, controlled with a gate and monitored from a guard's shack that was manned twenty-four hours a day, seven days a week.

Then, along the western side of the property, there was a second driveway for those headed to the main office building, but there was a gate there, too. This one was activated by an electronic fob system.

There were cameras at all strategic points along the edge of the parking areas and throughout all of the facilities that blanketed the area in continuous coverage.

The front door of the office complex was secured with a magnetic lock that restricted access to the receptionist behind the reception desk in the main atrium. After-hours access to the building was controlled by an advanced keycard lock system. Any breach of the lock would trigger a pulse alarm on the property and immediately notify both the local law enforcement and a private security provider.

Access from one department to another was again controlled by the use of keycards in order to prevent any unauthorized access from one suite of offices to another. And cameras were placed throughout the building to record all of the comings and goings within the hallways.

No, there was no shortage of security in place at the Carter Supply compound, but none of these features could stop what happened in the darkest hours of that midwinter night.

Despite all the safeguards, disaster slipped right past all of the locks without sounding the alarms and moved through the office corridors without hesitation. If any of the cameras captured an image of the shadowy figure as it went about its nefarious plan, no one took any special notice of those incriminating frames.

Sometimes the greatest danger a business faces can't be stopped by even the strongest lock or captured on the most advanced security camera. Sometimes the biggest threats to a company's success are far more insidious than that.

TVGENIUS

Have You Defined Your Differentiator?

This novel begins with the mystery of a perceived systems disaster impacting a company's ability to do business. Developing solutions in situations that initially feel mysterious is a skill that fits right in the wheelhouse of our people's expertise.

When it comes to unraveling the mysteries of an ERP implementation, we've found the difference maker to be people. When people ask me what makes The Vested Group and our people special, I always reply that it's our special sauce, what we like to refer to as TVGenius. Here's the dictionary definition:

TVGenius [tee-vee-jeen-yuhs], noun:

That unique set of superpowers each one of us has *that makes us brilliant or amazing. Sometimes this is easy to identify, like coding skills, nunchuck skills, bowhunting skills, or an uncanny encyclopedic knowledge of saved search wildcards, and sometimes it's more elusive, like being able to defuse a tense interaction merely by your tone of voice.*

That spark or flash of inspiration that pops up unintentionally and inexplicably wows everyone *on your project team and beyond and consistently underscores who you are and why you fit in at The Vested Group so well.*

The only accurate answer to the questions, Where did he learn that? How'd she come up with that? When did you find that? How did he know that? How did she do that? ***It's that cool thing you do that can't be learned or taught, but just is.***

Throughout this book and pretty much all of The Vested Group's messaging, you will see references to TVGenius. That's because everyone here has strokes of TVGenius, and all of these qualities can't possibly be defined by one real word.

As you read this book and explore The Vested Group beyond these pages, keep your eye out for those explicit and implicit mentions of TVGenius if you want to know what it is that makes us worth getting to know, amazing at what we do, and great to do business with.

NEXT STEPS:

Your company's message will flourish when you explore and define what makes you different. The Vested Group found guidance through the Rockefeller Habits and professional business coaching—this was a game changer for us! The Rockefeller Habits are intended to narrow your business focus and can be summed up by defining a few key company priorities, using a company-wide critical metric to monitor performance, and establishing consistent meeting rhythms or huddles. The Daily Huddle is an example of a Rockefeller Habit that was extremely impactful for us—everyone at The Vested Group participates in a twelve- to fourteen-minute huddle to start each day.

To learn more, I highly recommend the following:

Scaling Up, by Verne Harnish

Titan, by Ron Chernow

You can find a certified Scaling Up coach at www.gazelles.com. The Vested Group can confidently recommend coaching services from Petra Coaching or Michael Mirau.

CHAPTER 2:

STEVE AND THE VALLEY OF DESPAIR

STEVE E. NICHOLSON had spent most of his adult life in the digital business world.

Steve's career started with some of the biggest and most prestigious firms in the country, the sort of old and well-established groups that were not only known and respected within the higher corporate echelons but were household names as well.

Those professional experiences had all been meaningful, but there was something about those behemoths' approach to servicing their clients that failed to satisfy Steve on a personal level. That shortcoming certainly wasn't a result of any professional inadequacies on the part of his employers. No—to the contrary, they were all leaders in the industry for very good reason. And Steve was always aware of (and grateful for) the extraordinary professional opportunities that these positions had provided him.

Still, somehow at the end of the day, Steve felt that for all of their well-deserved prominence, those consulting giants failed to provide the sort of personal touches that were necessary to developing deep, meaningful client relationships in the way he thought they should exist.

Rather than having clients who were little more than interchangeable files that came and went without any sort of real human interaction, Steve was convinced there was a better way. He could deliver those same technically excellent skills but within the framework of a genuine consultant-client relationship that would afford him the opportunity to get to know his clients and their businesses while also maintaining a culture that would challenge and inspire people to be the best versions of themselves. This would reduce employee turnover and project churn, thus offering clients a level of service utterly unmatched in the industry.

> This would reduce employee turnover and project churn, thus offering clients a level of service utterly unmatched in the industry.

And so, Steve had left his prestigious job with the nationally known consultants and taken the bold step of creating a group all his own.

This was not a professional decision that everyone in Steve's life automatically understood or approved of. At least, not at first.

There was, of course, no shortage of consulting firms throughout the Dallas-Fort Worth metroplex—or throughout the country, for that matter.

In addition to the "big boys" that Steve had worked for, there were any number of regional copycats that offered the same type of services at a significantly reduced rate.

Steve, however, was quick to point out that these clones simply followed the industry leaders. The vast majority of industry players still possessed the exact same mindset that limited client contact throughout the sales cycle and implementation.

In fact, the one commonality that Steve found in all of the many options and offerings that were available in the market was that none of them offered the type of personal service that he would want if he ran a business in need of help in the assessment and upgrade of existing software. Not enough people were educating their clients on questions they should be asking. It's impossible to find the correct answers if you are not asking the right questions. "How will my people be trained before the implementation? What happens after the implementation? Whom do I contact for support?" Few firms covered these essentials.

And that was Steve's objective from the very beginning: to create the sort of consulting group that *he* would want to work for as an employee and to partner with as a business owner.

It had been more than twenty years since Steve founded TCG, and after all those years, he had learned that there were certain fundamental laws, and among them there were none any truer than this: the later the hour of the client call, the more difficult it was to reach an end-of-the-day solution.

And this particular day had already been over for some time now.

So, when his phone rang, Steve looked at his watch and saw it was 6:23 p.m. In that instant, he knew it was going to be a heckuva call.

He answered it like it was the first call of the morning anyway. "Good evening, Mr. Mack!"

Dean Mack was the CEO of Carter Supply, a manufacturing conglomerate located right in the sweet spot of North Texas. Why someone named Mack owns a business named Carter is an interesting story. In true Wild West fashion, the original Mr. Mack lost a bet that forced him to retain the Carter name upon purchase of a small subsidiary, and since he was a man of his word, he never even considered changing it back to the original name of Lucidity Metals.

Usually cool and controlled, his voice sounded nervously excitable. "Trouble, Nicholson. We've got all kinds of trouble."

"What's wrong, Mr. Mack?"

"Wrong? The whole thing is wrong! That's what's wrong."

Steve got to his feet and attempted to get the conversation on track. "Mr. Mack, help me understand the trouble."

"Nicholson, you promised me an ERP that would allow us to integrate our departments together into one system. We'd be able to track our resources, manage our business commitments, and share data in a way that would make all of Carter Supply more efficient in our operations, short term and over the long haul."

"I did say that," Steve said, "and it continues to be a significant portion of the project charter we put in place at Carter Supply."

"Well, now that we've gone live," said Mr. Mack, "all of my people are telling me they can't get their ordinary workload done. Everything is too confusing and complicated."

Steve nodded to himself. "Well, I think what you may be experiencing is the Valley of Despair."

"That doesn't sound good."

"It's normal," said Steve, "and temporary. A place we pass through as we all get up to speed on the new system. As your team will recall during our presentation a few months ago, we had a good discussion about the Valley of Despair. We even had a couple laughs, as your team anticipated the change an ERP implementation was going to bring. Nobody likes change. As we go from the first mountain, which is your previous system, to the next mountain, your ERP system, we have to cross the Valley. Thankfully, we're almost there. But it sounds like we still have some climbing to do."

"I recall the presentation and the laughs. But my people are struggling with the climb."

"We'll get to the other side," Steve said. "Remember, we discussed with your team that converting your operations to a new ERP was an invasive procedure. There's a learning curve that users are going to have to go through before they get completely comfortable with the new system. Give your team a few weeks and they won't even remember the name of the old system. I'm betting that's the issue at hand."

"Well, that's not what my department heads are telling me," Mr. Mack said, although his voice had relaxed with Steve's explanation.

"Well, what exactly are your department heads telling you?" Steve asked.

"They say they can't do the work because everything's wrong."

Steve nodded. "Mr. Mack," he said, "I promise we will continue to diligently work any system issues to completion. We'll resolve this with your team, ASAP. I can assure you that with a bit more looking into, we'll have every question sorted out. I'll be there first thing in the morning."

"All right," Mr. Mack agreed. "If today was any indicator of what tomorrow is going to bring, then the earlier the better."

"I understand," Steve said. "Good night."

"Good night, Steve."

Steve put the receiver back in the cradle by his desk and took a deep breath to collect his thoughts—a brief mental review of the day and what he still wanted to accomplish before he started again in the morning. He gathered his computer and the folders he wanted to review later that night and put them into his bag.

Down the hall, he found Maria, TCG's office manager and unofficial spark plug. She also happened to be Mrs. Steven E. Nicholson. At that moment, she was looking quizzically at her computer screen.

"Hey!" Steve said. "Are you ready to go?"

"Yeah, perfect timing," she said. "I just got all the paperwork sorted out with the contractors at the new office."

Together they walked to the back of the building and then to the parking lot beyond, where their car waited for them all alone.

Steve might have been the founder of TCG, but Maria had played a vital role in the company from the very beginning. She was far more than just the office manager.

If TCG were a living being—and in many ways it was—then she was the company's heart and soul.

She was mindful of all the things that might otherwise have gone overlooked in the occasional organized chaos that their business could create. She read the subtleties of their organization and exercised a style that allowed her to shepherd a collection of distinctly individual personalities within the growing number of team members in an effortless way that brought out the best in each and every one of them.

"Were you just on the phone with Mr. Mack from Carter Supply?" she asked as they got into their car.

Steve put his bag on the back seat. "Yeah, his people are having some troubles with everything as they move forward."

This wasn't her first rodeo—or ERP implementation. "The Valley of Despair?" she asked casually.

"I'm almost positive that's the case," he said. "As you know, the Carter Supply implementation was a project rescue for us. So, whenever we undertake a client that had previously been dealing with another firm and put together a plan to remedy that troubled project, we know the situation will be unique and will require a modified approach. I feel confident in our team, but you can't always predict an outlier."

"You know we'll always do the right thing," she said with a smile.

18

"True," he said. "But there's a little something extra working when we're rescuing a company from someone else's project gone wrong. Still, I can't see any reason there'd be any fallout with this implementation outside of the usual cutover areas, so it could be Occam's razor—"

"The simplest answer is usually the right one."

"Yes indeed."

"What are you going to do about it?" she asked.

"Well, I'll meet with him in the morning," Steve said. "We'll talk through the situation and get our arms around it. I think once everything is on the table, we'll find that the solution is a simple one."

She smiled. "Thank you, Mr. Occam." She kissed him on the nose and turned the key. "Better take Hannah with you tomorrow morning."

TVGENIUS

Should You Look for an Industry Expert or an Implementation Expert?

Choosing the right partner for your implementation is easily one of the most important factors in determining the success or failure of the overall journey from your previous system to the new one. Everyone knows that the right partner makes all the difference, but defining a successful partnership can be a bit more elusive.

I'll make it easier for you. Here are the top two must-haves to look for in your implementation's other half: expertise and trust.

First, a great partner recognizes that you are the expert in your business, even if you are most likely not an expert in ERP implementation. When you choose the right companion, they will bridge the gap between your business expertise, their experience in companies of similar size and scope, and their expertise in configuring your ERP system.

The expertise of both partners is essential for a successful outcome. But expertise only gets you so far.

That's why the second must-have for an implementation partner is trust. A great partner realizes that your ERP implementation is a huge investment for your organization and that you have understandably high expectations for what it will do for your business. The right partner will ensure you are comfortable with your system configuration as well as the rationale behind the decisions made along your implementation journey. You may not always agree with the recommendation, and you may even have some spirited conversation about the topic, but the decision is always yours to make. You should expect your partner to challenge your way of thinking and question existing legacy processes. If your partner ever asks you to tell them what you want them to do, run. It's their job to drive the process, but it's also important to let them. If they earn your trust, then you should trust them.

A great partner will want to trust you, too. They should want to understand and appreciate how your company culture will respond to challenges presented. How do you believe your team will react to stress and difficult decision making? Will people support or undermine each other? Do you have leaders that will guide and inspire?

Choosing the right business management software is an early step in preparing to scale an organization. Choosing the right partner—an experienced one you can trust—ensures your growth will be successful.

NEXT STEPS:

It's easier to choose the right partner when you know which questions to ask. Please visit joelpatterson.com/the-book to view the partner selection checklist.

CHAPTER 3:

HANNAH AND
THE CHANGE GUIDES

TCG WASN'T THE first consulting group that Steve had founded. When he first set out on his own, Steve had created a consulting group that he thought would be the ultimate expression of both his strategic approach to enhancing businesses through leading-edge technology and his own personal core values.

The endeavor was a success.

In fact, if such a thing were possible, it was too much of a success.

The company grew and grew and added consultant after consultant to accommodate and perpetuate that growth.

From a certain perspective, it was a terrific result.

From Steve's perspective, however, he ultimately began to feel as if he'd simply traded the coldness of the international "big boys" he'd left behind for the same impersonal style of his own rapidly expanding consulting group.

And so, Steve had simply started over.

This time around, however, he had concentrated on an unwavering commitment to being people centric, cultivating a culture that embraced its core values on a very real level.

Each member of the TCG team was an expert in their own right, but they had been hired by Steve because they had shown some additional personal qualities that allowed them to share his vision of building relationships with their clients—and one another.

Hannah, the lead on the Carter Supply implementation, was no exception.

Hannah had been with TCG almost since its creation, and if Steve had a blueprint in mind for the perfect employee, someone who had an expert knowledge of the technical elements in which they worked and was also a consummate teammate to all coworkers within the group, then that looked an awful lot like Hannah.

The ride out to Carter Supply was like so many other trips around the metroplex: there were lots of ways to get there, but no good way to go.

Still, Steve thought the standstill traffic on I-75 was a perfect opportunity to discuss the Carter Supply implementation with his passenger, Hannah.

"What's wrong with the Carter Supply installation, anyway?" she asked.

"Do you think maybe it's a case of everything is new rather than everything is wrong?"

Steve's answer was simple enough. "The nature of their problem is everything." Steve shook his head. "It was after hours, and Mr. Mack didn't say much other than that his department heads are stuck, so I can't say for sure. And I'm not even entirely sure that he knows, yet. The system has only been in place for a couple of days now. But the folks at Carter Supply seem to think that *everything* is wrong."

"Oh," Hannah said, with a knowing shake of the head. "Well,

those first days of adjustment can be trying. Do you think maybe it's a case of everything is *new* rather than everything is wrong?"

"That's my hunch."

Hannah nodded. "When you think about it, the implementation of an ERP is so significant that there are bound to be adjustment issues, especially in a company as big and as diverse as Carter Supply. That's why we train our clients throughout each step of the process. Then, once we go live, it's our aftercare that sets us apart from any other firm."

"And that's why," agreed Steve, "we stress how important it is that our clients come to a project not just as clients but as partners. For every implementation step we take, I think the most important thing a client can do is to take a matching step, by ensuring their people are kept up to date on the progress of the project and that all of their people are adequately prepared and trained when the system goes live."

"Especially when it's a project rescue, right?" added Hannah. "Those are as intensive for clients as they are for us. Perhaps now the Carter Supply team is facing the reality of the new system and finds it to be more encompassing than they anticipated in the beginning. Thankfully, we were really thorough on all of those preparations and designs—even for us. I can't imagine that it will be too difficult to fix."

Steve nodded. "I think that's right."

"It was a complicated scenario," Hannah added as an afterthought.

"Sure was. They're a manufacturing company, but they'd acquired some other enterprises that had to be worked in at the last minute."

"Change is hard for everyone," Hannah said. "Just renovating the new offices and moving certainly has its set of challenges even for us."

"Don't I know that's true."

"That's why it's so important that our clients embrace the commitment to move forward …" Hannah trailed off.

Steve looked over at her and could see something was on her

mind. "What is it?"

Hannah grinned. "I was just thinking of the new Green Legacy account. With David Martinsson."

"Oh yeah? What about it?"

"Just that Green Legacy has operations almost as complicated as those of Carter Supply. I hadn't realized it until now, but I think it'll be interesting to see how we close out this implementation with Carter Supply. It'll be good to apply what we learn to the new project with Martinsson."

"Good thinking," said Steve. "What makes the Green Legacy operations complicated?"

"Mostly that they're a green manufacturer and refurbisher that has been in business over fifteen years, and they're looking to maintain a carbon-neutral environmental impact. This means their supply chain has to account for more variables than those of most companies. And more regulations. Plus, over the past year, Martinsson has acquired additional sustainability-focused organizations that refurbish almost anything you can imagine. So now their newly combined operations and finance teams are as complicated as their vendor base."

"It's funny," Steve said, "that even with all those unique variables, the most important part of our job will be training the people."

"That's for sure," Hannah said. "Just like with Carter Supply. You've got all of these employees who've been using a certain system and doing things one set way, over and over again, for five, ten, fifteen, maybe even twenty years. For their whole careers, they've been doing whatever they do one way, and then all of a sudden they're tasked with making a big change. Then they invest time and energy in the first firm's process and methodology, which doesn't get off the ground. By now, we've come a long way in training, but the Carter Supply team had a big hurdle to overcome at the starting line."

"Even though the goal of the system we implement is to make things simpler for the team in the long run," Steve said, "sometimes it's hard to see that big picture in the short term."

"That's why," Hannah added, "what we offer our clients is *more* than just new systems and the services to support them. It's a partnership. That has always been our goal at TCG; it's how we want to distinguish our consulting group from the others. Our role has to be guiding our clients and their employees through what you and I know is often a challenging transition."

They pulled up to the gate entrance to Carter Supply.

"I suppose that's the odd thing about what we do," Hannah said.

"What's that?"

"We help people improve their business, but that makes us agents of change. And since change is uncomfortable, I guess not everyone sees us as the rescuing-hero types."

Steve nodded. "I suppose that some see us as the people who are turning the old and comfortable into the new and challenging."

"And that explains a lot," Hannah said. "Like why our clients think that something has gone wrong with their project when it's just a few outlier hiccups shading the client's perception. It's almost always the case that things are going well—or at least as well as expected— but the client is struggling with a new system or new processes that require employees to do whatever it is that they do a little differently than they're used to doing it. So, it's not that the system isn't working; it's that they see *different* as the same thing as *broken*. Do you know what I mean?"

"I do, and once again your wordsmithing is top notch."

"Guiding them through change," Hannah said, suddenly smiling. "Are we 'the Change Guides'?" she asked, playing a game many play at TCG: guessing what the company's initials stand for.

"The Change Guides? Something like that," said Steve, who liked to play along, giving the impression that the firm's name was in fact a mystery.

"The Change Guides." She repeated the words like a voice actor reading a title card. "We should so totally be superheroes or something."

As Steve pulled into a parking space, Hannah looked at the place with fresh eyes. "It's funny to think," Hannah said. "I remember the very first meeting with everyone at Carter Supply. And now here we are on the other side of the implementation."

"That's how it goes," Steve said. "We're there from the very beginning until always. Now it's important to let Dean and the rest of the folks here at Carter Supply know that we're going to stand by them to sort this out."

"It certainly takes a big commitment from both sides to make the change," Hannah said.

"The Change Guides and the Big Commitment." Steve looked to Hannah. "Kinda catchy. What do you think?"

"I think Maria is an angel."

He smiled at himself. "Yeah, she is."

TVGENIUS
Have You Been to the Valley of Despair?

At first glance, the Valley of Despair sounds like a dark, scary place where Negative Nancy likes to go camping. Thankfully, this is not the case. The Valley of Despair is actually a place of hope—a place you find yourself when you are learning something new, your frustration level is at its peak, and you just want to give up. But you don't give up. With each day, your learning curve shrinks as your competence with the new system grows, and before you know it, you are out of the valley. Simply being aware that the Valley of Despair exists, bracing yourself for the dip, and knowing that you will come out on top is the key to success.

You might assume that you will find yourself in this valley at go-live when you flip the switch and your new system is up and running. In reality, the actual day of go-live is pretty anticlimactic. It's typically pretty quiet and uneventful. Keep in mind that the go-live date isn't chosen as the point your system is expected to be in a state of perfection. Given the nature of this journey, perfection isn't really possible anyway. Go-live is more about finding the tipping point of when you're 90–95 percent sure nothing major is going to blow up—you're able to ship product and bill customers, but small issues are going to happen. It's the days immediately following go-live, when transactions are rolling through and users are uncovering the hiccups and sidesteps of their new system—not to mention adjusting to the process changes they are getting acclimated to!—and it feels like it's one break/fix or training session after another and it is easy to get overwhelmed and bogged down by the struggle.

We see this happen at every single client site, and the reason I bring up the Valley of Despair in our kickoff meeting is not only so the client can brace for it but also to normalize these feelings. Yes, you will hit your lowest low and feel defeat, but you are not defeated. As the days after go-live progress, errors will become fewer, reports will make more sense, your comfort and confidence will grow, and you'll begin to see things trending upward. Before you know it, the system will be perfunctory—dare I say boring!—and that's when you know you are ready for Phase Two …

Just wanted to point out another facet of TVGenius that we bring to the table: the agents of change *principle. Being successful agents of change starts with knowing how to put a project team together. Your internal project team can make or break the project, and to set you up for success, we make several recommendations. Two important points we make are as follows:*

Your project team must include people who make decisions and have the will to implement them. Projects are about velocity and momentum. Decisions that drag out multiple days have a major downstream impact. A day lost here and there turns into a missed go-live date before you know it. You need people on your project team who have the authority and wherewithal to keep the project moving forward.

Successful project teams encompass a cross section of team members spanning your organizational chart from senior leadership to department functional owner. Having this cross section helps you avoid the pitfall of ending up with the exact same processes you have today but applied to a brand-new system. The traditional, old way of doing things may work well, but the check and balance of new eyes and fresh ideas is invaluable. Having old, new, high, low, and middle come together will give you the best possible perspective to inform better business practices.

NEXT STEPS:

Believe it or not, you can have fun with the Valley of Despair! Print out a copy, post it in your conference room, and let your team track their journey through the valley by moving their pin through the stages. Creatively reinforce that struggling is normal but that you will make it through as a team.

Please visit joelpatterson.com/the-book to learn more about the Valley of Despair!

CHAPTER 4:

MYSTERY STRIKES AGAIN

SOMETIMES THE DARKEST of deeds happen in the dead of night when there's no one there to witness them.

Sometimes, however, the act is even more sinister still because it happens during the light of day and right under the nose of any number of unsuspecting people who simply had no idea that someone within their midst was carrying out some malicious intent.

The morning at Carter Supply was seemingly like any other.

A slow but steady line of employees filing into their offices for another workday. Folks settling into meeting rooms and desks or chatting over coffee in the breakroom.

Two floors away from the breakroom, a figure crept down the hall, trying hard to look as if they were just another part of that all-too-average morning and that there was no ill intent on their mind.

A disarming smile to those who passed by in the hall.

A friendly nod of the head as if everything was business as usual.

And when the coast was clear and no one was looking, a quick swipe of the key lock.

When the door closed again, there were no witnesses.

And all those in their offices and cubicles never suspected a thing.

None of them had the slightest suspicion that it was not just another ordinary day.

Not anymore.

CHAPTER 5:

MR. MACK

THE FOUNDATIONS FOR Carter Supply had been set on a forty-acre patch of North Texas dirt and scrub grass long, long before Dallas's unchecked development had spread down the I-35 corridor and surrounded it with pop-up condos and strip malls. In fact, long before I-35 itself.

Though it now saw annual revenues in excess of $100 million, the conglomerate had started in the wake of the Dust Bowl as just another scrap pile: the ol' Mack Junk Yard.

Despite the economic devastations of the era, every misfortune hides within it the seeds of opportunity. The original Mr. Mack—Daniel, or Dan to his friends—had realized there was a small fortune to be made in selling the metal scraps he'd collected during the Depression to feed American manufacturing during WWII.

At the height of the war, Dan Mack took that initial inspiration and his life savings to construct a small smelting facility in the back forty of the junkyard, and that was the humble beginning of the organization that ultimately became Carter Supply. He had started the process of converting the metal himself, melting down the junk he collected and manufacturing sprockets that turned out to have a thousand different uses in the Allied war effort.

Prosperity held even more possibilities, and in the golden days of the post–World War II economy, the Macks' industrial empire had grown and grown until Carter Supply was known as the country's premier sprocket manufacturer.

American-made since 1938, as they liked to say.

The old plant still sat on those original acres, but it was now dwarfed by the addition that Daniel's son, Matthew, had built onto it in the Reagan boom days of the 1980s.

A little time after that, Matt built a warehouse big enough to house both supplies and manufactured product.

Next came a distribution facility on the property, with parking for a fleet of twenty tractor trailers, all of them with the Lollar Transportation Company name (another lost bet—Mr. Mack was a great businessman but terrible with odds) across the cab door in bold red letters. Their primary function was to deliver manufactured product, but when those deliveries from Carter Supply were completed, they carried an assortment of freight for other companies as a third-party contractor rather than make their return trips empty.

And then at the beginning of the new millennium, Dean Mack— the third generation of Macks to run the company—built Carter Supply Corporate Center.

He built his office building with the idea of capturing all of that high-rise glamour just off to the northern horizon but still staying true to the company's roots and original location.

Inside the building, Steve and Hannah stood in a more-than-comfortable waiting area just outside the executive offices.

"Isn't this nice?" she whispered to Steve, who hadn't been in this section of the building since early in the implementation.

"It is," Steve agreed.

"I could just move my apartment in here," Hannah joked.

"Wait until you see our new office," Steve said.

The doors parted, and out stepped the impeccably dressed Mr. Dean Mack. "Good morning, everyone."

Steve and Hannah got to their feet.

"Good to see you, Dean," Steve said, shaking hands.

Mack extended his hand to Hannah. "Good to see you, Hannah."

"I wish it were under better circumstances," she said.

"Just what are the circumstances?" Steve asked.

Dean checked his watch. "It's been a heckuva day already and I'm running late, but let's take a minute to sit and talk this over."

Dean showed Steve and Hannah into an office much nicer than the luxurious waiting room.

Steve and Hannah took the leather chairs they were offered, while Dean Mack made himself comfortable behind his desk.

Dean took a deep breath and sighed. "I have to tell you, Steve, I didn't expect any of this when we started down this road."

"I understand," Steve said. "We talked about the importance of managing expectations when we started that journey, but let's just take a moment to discuss specifics here."

Dean nodded. "I'm having all of that put together ASAP, so all I have right now are complaints. And they are starting to pile up."

"Complaints from your department heads? Or from the users?" Steve tried to clarify.

"From department heads," Dean said. "Or rather via department heads." He took a moment to get his own thoughts straight. "I suppose the complaints start with the users, but I had department heads on the phone and in my office all day yesterday."

"Okay," Steve said. "What are they complaining about?"

"I've asked everyone to put their complaints in a report we can get down and work with, but yesterday all they had for me were generalized complaints that no one could get their work done."

"But no specific reason why?" Hannah asked.

Dean gestured with his hands. "The only specifics we have are that we're behind where we should be, and that's just not a hole that I can afford to allow this company to fall into—especially not with everything that we went through with that first implementation group."

Steve knew that the first rule for any fire—be that forest or house or client—was containment. "All right, then the most important thing we can do right now is assure you that we're confident with our design, our work, and our installation. We don't have any one-off clients. You're a long-term partner to us, and we're going to be here and do what it takes to resolve the issues you're encountering."

"I know that, Steve. And we feel the same way about you. But we need to get this worked out as quickly as possible, because every day we're down—for whatever reason—is one more day than I can tolerate."

"Mr. Mack," Hannah said. "I have to tell you that the fix may be as simple as redoubling your efforts to refit and train your employees. There's no doubt that the implementation is a major undertaking and by necessity it can create a disruption in the ordinary course of business. It is, after all, a completely new system, and this requires that everyone within your business essentially relearn their jobs and potentially perform the same operations they've likely been performing for years in an entirely new and unfamiliar way."

"I won't turn down more training," said Mr. Mack, "but my understanding is that the report will show issues beyond training."

"That would surprise us, Mr. Mack," said Steve, "but we'll remedy any and every issue."

Dean Mack checked his watch again. "Listen, I wish I were able to stay here with you and work through this situation, but I'm afraid I've got a videoconference with a supplier that I just have to hop on right now."

Steve and Hannah got to their feet.

"Of course," Steve said.

"Listen," Dean said. "Why don't you stop in and see Ben. I think that he could probably point you in the right direction, open the doors that need to be opened for you to have the look-around you're going to need to figure out this problem of ours. After all, he's the guy who brought you to us in the first place."

Steve nodded. "We appreciate that."

Dean shook his head. "I only wish that he'd done it sooner. That group we had in here before you—they were go-getters, but they just weren't up to a task of this size. Hiring them was a joint decision, but I've been in business long enough to know that it's always foolish to shop based just on price." He shook his head. "I should have known better."

"I wish I could talk to every company out there that is preparing to make this upgrade to their systems," said Steve. "Not to sell them, but just to advise them how important it is to move forward with the right people who are truly invested in the client's success and long-term relationship. Because what we really want to do is forge a relationship with the client in which we're no longer just consultants and they're no longer just consumers of our services. Instead, we're *partners* working together to better their business moving forward. That's a delicate relationship, and it requires everybody to come to the table with an open mind and a shared commitment."

"Well, I tell you what. You fix this problem, and we'll have all the partnership you want."

TVGENIUS

Is the Price **Really** Right?

Once you've evaluated your situation and made the decision that you need to move forward with implementing an ERP system, your sole focus should be on finding an implementation partner that has an established track record, has verifiable recommendations from former clients, and—and this is the most *important factor—proves to be a good match with your people. You can flip back to the TVGenius section in chapter 2 to find a valuable Partner Selection Checklist. Whatever decision you make, please don't make the decision based on price alone. Your business is far too important.*

My advice is to initially make this decision without any consideration of price and then use price as a data point for the final negotiation.

I say this not only with regard to those who might otherwise be driven to save money but also to those who think that the highest bid on the table necessarily represents the best qualified team to serve your unique business interests.

If you take nothing else from this book, you could do worse than to come away with the resolve that finding an implementation partner is a people issue, not a monetary one. It is rare for any client to enjoy experiencing the changes that come with a software implementation. It doesn't matter which software you've chosen or whom you've selected as your implementation partner—nothing and no one can make change painless.

The TVGenius in our approach to change management boils down to honesty. We are not afraid to ask the tough questions and shoot straight with our answers. If anyone tells you that implementing a new business management system is easy or without pain, they are lying. There's no middle ground here. The sheer number of processes to be reviewed, master data files to be scrubbed, AR invoice formats to be ridiculed, and people to train will likely impact aspects of your company you didn't think were in scope.

The goal of your implementation is to have a system that runs your business efficiently and with reliable data. What that looks like and the path to get there are different for every client. There's no recipe, per se. You have the list of

> Our approach to change management boils down to honesty. We are not afraid to ask the tough questions and shoot straight with our answers.

ingredients and a general flow for baking, but the oven temperature is impacted by external factors like humidity and elevation. We understand that you've built your business processes over many years—change and improvement won't happen overnight.

So, when it comes to change, we sometimes see that clients have a tendency to develop unreasonable expectations. That is not to say they have unreasonable expectations for the performance of the system or for our level of support and follow-up. No, what they are unrealistic about is their own "change appetite": their ability to accept and adapt to the significant amount of change that the implementation will undoubtedly (and unavoidably) bring to their business.

Your implementation requires that all of your employees somewhat suddenly learn all new policies and procedures. What we're talking about here is leaving behind a comfortable level of mastery of an old system and learning a new skill, a new system. Our aim is not to eradicate the pain of change but rather be a supportive partner in guiding you through the transition to a better way of running your business. Think of your implementation as building a house you intend to live in for twenty *years. How would you think about the foundation or roof differently? Let's make sure the foundation can withstand the expansion and contraction of Texas soil first and worry about the pool and second garage later.*

NEXT STEPS:

It's always reassuring to get a feel for what it's like to work with an organization before you actually sign on the dotted line. Asking for a sample document is a great way to get a feel for whom you are working with. As an example, you may take a look at a way to identify project risk that The Vested Group uses. You can learn more at joelpatterson.com/the-book.

CHAPTER 6:

BEN

"WE'RE STILL SMARTING about the fact that we shopped on sticker price," Ben Mixon said, shaking his head. "The first firm came in at the lowest price but then shut down our whole operation for a while. By the time we realized they couldn't execute, it was too late, and we had to revert back to our old homegrown system. Killed our fourth quarter."

Ben sighed. As the chief financial officer for Carter Supply, it was in this capacity that he had first reached out to Steve and the folks at TCG to take over an installation that had gone very, very wrong.

"You're absolutely right," said Steve, who along with Hannah had made the short walk over to Ben's office as soon as they'd adjourned with Mr. Mack. "When you're considering something as intensive and invasive to your business as an implementation of this magnitude, price is clearly important but should be one of the last considerations to factor in. After all, in the end the lower price the other group offered you turned out to be the most expensive route to take."

Ben nodded. "You don't have to tell me. When we initially started to consider upgrading our system, my first instinct was to turn to the established consulting groups—you know, the usual suspects. But there were concerns about costs. We're a fair-sized company, but we've got a lot of very unique elements. I knew we weren't the sort of company that would be well suited to the big consulting firms that treat every client like they're exactly the same. And more than that, I knew the price tag on dealing with those major players was going to be in the stratosphere. There was no way Dean was going to pull the trigger on something like that, not with some of the numbers those people were floating past us."

Steve understood the situation perfectly. "Of course, going in the opposite direction didn't prove to be the best answer, either."

"That other consulting group was simply underqualified to handle an implementation like this," Hannah added.

"It's true," Ben acknowledged. "Everyone on the committee agreed with the decision, thinking we were going to be okay pushing the envelope by taking the very lowest bid. Ultimately, we paid the price for cutting those corners."

"Well, at least we can all say we did our best to get back on track once we got engaged with your team," said Steve.

Ben nodded. "We all did our best, sure. But it's been tougher on some than on others. Brittany, for example. She volunteered to lead training—"

"And did a great job," Hannah interjected.

"That's right," said Ben. "I only wish it didn't prove to be such a big task. Because of the flop with the first firm, Brittany really has led training for *two* implementations instead of just the one. And because we had emotionally invested in the first firm and then were let down, I think we were all a bit drained before we started your training. I even noticed that our overall attitudes, office wide, seemed to shift the week we realized the first firm's training wouldn't apply."

"Well, we're here to remedy the situation. We won't let you down."

"Don't I know it. I fought to bring you guys in because I knew you had the right experience and team for the job. I only wish we'd brought you in sooner rather than later."

"Me too," Steve said. "If you had consulted with us initially, we would have walked you through the planning stages and explained to you how important every step is right from the beginning. I mean, we did it for you with our remediation plan, but in our Phase Zero process, before we start any implementation, we've thoroughly reviewed a client's three to five critical business processes, not to understand their business as well as the client does but to understand how and why they arrived at the point they are now at. Then we work out a detailed plan that covers every aspect of the implementation so that there are no misunderstandings about what is going to happen and so that we can anticipate potential problems before you have a situation like your production line going down for thirty days. Following our proven Phase Zero process mitigates these risks and allows our team to focus on the areas that truly matter to the business."

"Well, all we can do now is make sure we fix what's ailing us," Ben said.

"So," Hannah asked, "what exactly is ailing you?"

Ben nodded. "What I know is that we all started the week like it was just a regular work day, but it was anything but. About four o'clock I got a call from Dean, who said that everyone was telling him it doesn't work."

"That seems to be the consensus," Steve said. "That the department heads were complaining that their people couldn't work with the system. Mr. Mack said he'd asked for reports on the *specifics* of the problem. Have you seen any of those yet?"

"No, I haven't seen anything like that yet. Of course, giving them their due, it's still early, but I don't have the exact answers I know we need," Ben said.

"So, what do you see as our first steps here, Ben?" Steve asked.

"You'll have to ask a couple of department heads to get the answers you need."

"And where would be the best place to start asking?" Hannah wondered aloud. "With Brittany?"

Ben thought the question over. "Brittany is out today. So I suppose you start with Craig and Walker. In the change committee, they were the first to make the case to keep the status quo. After we all agreed, they made a solid effort to adapt when the first firm started up. But as things went bad, they've started pushing to return to the old system. I don't blame them, as the changes have affected their departments the heaviest."

"All right, we'll talk to those guys and figure out what went wrong, or at least get them to point us in the direction of some answers. That's step one.

"Then, for step two," Hannah said, "we'll go back and take another look at each planning stage of the implementation."

"Thanks," Ben said, reaching out and shaking both their hands. "I'll check on the report and send it as soon as possible. In the meantime, I really do think that you should talk to Craig and Walker. They're the guys that are in the heart of all this."

Steve nodded. "That seems like as good a starting point as any."

Ben dashed off down the hall and left Steve and Hannah alone for a moment.

"Well, I suppose we should go look in on Craig and Walker."

Hannah nodded. "I think you're right about the problem here."

"You think they're all just sliding down into the Valley of Despair?"

"Sliding? They're at the bottom. Luckily the only way to go from here is up."

TVGENIUS
What Are Your Core Values?

Our six core values not only define who we are but also what we look for when adding to our team. We acknowledge and celebrate these qualities in each other on a daily basis. Whether you are a prospective client or a potential employee, you can get a quick read on who we are and where we stand by taking a look at The Vested Group's core values: "All In," "Own It," "Always Do the Right Thing," "Give Them Your Shirt," "Appreciate That Failure Results in Learning," and "Enjoy the Ride"!

One of our core values speaks to the way we feel about integrity: "Always Do the Right Thing." A large part of doing the right thing is not shying away from the tougher conversations with our clients. While we may have to deliver a message we know a client doesn't want to hear, we also know that our clients appreciate our willingness to do so in the long run. You may encounter partners or vendors who will do or say anything in hopes of pleasing their client. We've found that mutual respect and honesty—essential facets of "Always Do the Right Thing"—make for a stronger and more successful relationship. And really, at the end of the day, what do you appreciate the most—being told what you want to hear, or being told the truth?

NEXT STEPS:

If you haven't already, commit to defining your company's core values. You'll be surprised at how often you refer back to your core values, and defining what these are to you is an exercise you'll never regret completing. The Vested Group found a lot of value in using the Rockefeller Habits to help us develop our core values.

CHAPTER 7:

CRAIG AND WALKER

DOWN THE HALL from Ben Mixon's office, at the very end of a long corridor decorated with portraits of all those who had served Carter Supply as executives, were two offices that faced one another. Steve and Hannah first stopped in the office on the left.

"Walker," Steve said as he shook the hand of Walker Arnav, who was Carter Supply's chief information officer. "Good to see you."

"What am I? A second-class citizen?" the other man quipped.

Steve stepped over to Craig's office. "Good to see you, too, Craig," he said, shaking hands with Carter Supply's chief technology officer, Craig Myra.

"And, yes," Walker called out. "As an Oklahoma fan, you are most definitely second-class everything."

"And when was the last time the Longhorns won the Big 12?" Craig zinged back at his coworker.

"We've had a rebuilding year," Walker offered.

Craig let out a very fake laugh. "Rebuilding year? I think you mean rebuilding decade."

"Gentlemen," Hannah chimed in, "Ben said to stop by and see you two, as you could give us some specifics on the issues with the live system?"

Craig was the first to jump on that. "Sorry to say, but the problem is that the ERP system doesn't work. I've had my eye out for issues ever since we first formed the committee to consider the suggestion, but this new fiasco is more messed up than the Longhorns' offensive game plan."

"I think *I'm* the one who first raised the red flags," Walker countered. "And this system has more problems than a Bob Stoops locker room."

It was a step over the line for Craig. "You take that back! Coach Stoops is a saint!"

"Guys," Hannah interceded. "No one has told us yet what the problems are."

"The problems are across the whole company," Walker said.

This appeared to be the only thing Craig could agree with. "That's right. They're all across the company."

"Supply and distribution are all tangled up," Walker said.

"And I've had a steady stream of calls from sales saying that they can't do a thing with the system," Craig added.

"The old system we had in place worked just fine," Walker insisted. "There wasn't any need to make any changes back then, and now your '*improvements*'"—he used air quotes—"have set this company back I don't know how far."

"I told you this was going to happen," Craig said.

"Walker. Craig," Steve said. "Let's be honest with one another here. We all know that Carter Supply's old system was *way* overdue for this upgrade. What you had in place was just too twentieth century to keep up with twenty-first-century business demands much less maximize Carter Supply's substantial growth potential."

Walker looked at Craig.

Craig looked at Walker.

Craig decided to double down. "Lord knows that Walker and I don't agree on much, but he's absolutely right that there was nothing wrong with the system we had in place. It's the new system that is causing all of the headaches."

"Well, we haven't had the chance yet to determine that there are any problems," Steve said. "But the legacy system that you had been using *was* a problem. The fact is that the longer a business hesitates to pull the trigger and puts off making these modifications, the more difficult the process and the implementation become. That's just a given."

> The longer a business hesitates to pull the trigger and puts off making these modifications, the more difficult the process and the implementation become.

"We were doing business just fine," Craig insisted.

Hannah said, "Guys, the days of a business doing 'just fine' are long gone. The name of the game today is optimizing your functionality so that you can maximize results while remaining flexible enough to pivot the business as future requirements change. That's no longer aspirational; it's simply a matter of survival."

Neither Craig nor Walker had a response.

Craig looked at Walker.

Walker looked at Craig.

"Where do we begin?" Craig asked.

"There hasn't been time just yet to get everything together," Walker explained.

"Well, once you have it, we can help," Steve said. "We made clear from the get-go when we started the process that there will be a challenging period of transition here, so we just need you to help us help you navigate that territory."

"I don't know. Coach Stoops stepped down and Lincoln Riley took over, and my Sooners still went to the national playoffs," Craig offered. "Sometimes transitions go smoothly. Or, at least, more smoothly than this one."

"Craig, we need you to be Lincoln Riley in this case," Hannah said. "We need you to help everyone on your team step into this new system."

"I'm not agreeing with Craig," Walker made clear. "But the system just doesn't work."

Hannah nodded. "Nothing against the Sooners or the Longhorns, but sometimes when a team gets a new coach and installs a new system, the players are used to running plays the old way and they struggle to catch up with the new playbook. Or a new quarterback comes on board and it takes time for him to gel with his receivers and backs, for him to get the rhythm of the O-line. I think that's exactly what we're dealing with right now. We just need the game-day tapes so that we can see for ourselves."

"Oh, that can't be the case here," Craig said.

"Our people know what they're doing," Walker added.

That seemed to be another of the few things on which they both agreed.

Steve corrected them. "Well, they *knew* what they were doing on the old system. That's not any criticism of any one individual, but they need to get used to doing new things in a new way."

"That's my point: everything was working just fine," Craig said.

"Listen, guys," Steve said. "Hannah and I are going to go back to our offices and sit down with the project team to see if they are aware of any issues. I don't think they'll have anything new, but I want to check that out just to make absolutely sure. In the meantime, we'll look for the punch list of issues your users are having. Then we'll come back to work through that punch list, making sure everything is functioning perfectly in every department."

"We can do that."

"If you can help us coach your team through this transition, I promise it won't turn into a rebuilding year. More than that, I promise you absolutely everything you both oversee for Carter Supply will work better than you could ever expect.

> We're all committed to winning, but that only comes when we all work together.

You've just got to trust the process and let our team be part of your team. All I want is for the Carter Supply team to come out winning."

Hannah added, "We're all committed to winning, but that only comes when we all work together."

Both Craig's and Walker's faces lit up when they heard that. "All right," they said together.

"Great," Steve said. They all shook hands, and then Craig and Walker disappeared into their respective offices.

Steve and Hannah walked down the hall.

"Well, I think that went well," Hannah said.

"Yeah," Steve agreed. He offered his best comic book narration voice. "I think our heroes are well on their way to wrapping this up … the Change Guides will soon be out of the Valley of Despair."

She groaned good-naturedly at the dad joke. "Come on, I've got to get back to prepare for that meeting with Green Legacy."

TVGENIUS

Is the Client Always Right?

Client care is a priority to *The Vested Group, and we take pride in the level of service that we provide. At the same time, this doesn't mean we are willing to do anything a client wants, whenever they want, no matter what.*

The fact of the matter is that our clients hire us to be the experts during an implementation, and we wouldn't be doing our jobs or exercising our responsibilities if we simply caved in to every client demand. Ours is not a "the client is always right" shop, because that's obviously not always the case.

We create happy clients not only by taking care of and protecting our team but also by forming trusted relationships with our clients. That way, when a difficult situation occurs, we can steer them in a direction they might not necessarily want to go by educating them and explaining the situation in a professional and respectful manner.

And in those situations, we almost always find that they come around and are glad that they trusted us and our expertise in finding a solution to the problem at hand.

I have to say that we at The Vested Group are extremely proud of the many, many recommendations we've received from former and current clients. In any business, a sterling reputation is one of the most valuable assets that any enterprise can have, and we are beyond grateful that so many are willing to stake theirs on our team and the work that we perform.

NEXT STEPS:

They say hindsight is twenty-twenty, but there are ways to evaluate and mitigate risks before you make impactful decisions. For instance, a preimplementation risk assessment looks to avoid as many of the bumps in the road as possible. Additionally, a RACI chart can be a great tool for clarifying ownership, communication channels, and accountability. If you would like to view a sample preimplementation risk assessment or a sample RACI chart, please visit joelpatterson.com/the-book.

CHAPTER 8:

MYSTERY BEDLAM

THE EXPLANATION MIGHT have been as simple as biology. The human mind is, after all, hardwired for routine.

Each of us is designed to overlook all of the little things that diverge from the norm and concentrate instead on the things that are "supposed" to be in place.

In essence, human beings are engineered to see what they expect to see rather than take notice of what is actually taking place: a natural ocular handicap.

If there is something out of the ordinary, most people will simply look right past it as if it's not happening at all.

And this simple trick of human nature may be the very reason that no one in Carter Supply's offices took any particular notice of anything out of the ordinary that afternoon. The fact that all of those people were hard at work as part of their normal routine may be the only explanation necessary for why not a single one of them took notice of the figure that had no reason to be walking down that hallway.

Instead, every one of those people—potential witnesses all— simply went about their business without paying attention to anything or anyone at all.

There was, however, something most unusual taking place on that otherwise ordinary day.

And there were abundant clues that would have been readily apparent to anyone who had been mindful enough to look past the concealing curtain of conformity.

There was, for example, an office that was left empty without explanation when someone should have been at their desk.

Then there was that figure that walked down halls that should have been empty and appeared as an unexpected visitor in sections of the complex where they should not have been at the time.

And there were other clues as well, if anyone had noticed.

Still, no one noticed the figure as it made its way to carry out its nefarious mission.

And no one noticed as the guilty party sneaked all the way back as if nothing had happened at all—as if they had never been there in the first place.

Of course, they had been there.

And something had happened because of that.

Still, despite it all, the only thing anyone at Carter Supply noticed was that the new ERP system, the one that had been giving almost everyone a degree of difficulty, had suddenly begun performing in an even more frustrating manner.

Things had gone from bad to much, much worse.

There were reasons, of course, but nobody noticed those at all.

PART TWO

THE
PLOT
THICKENS

CHAPTER 9

PROJECT MEETING

STEVE HAD CERTAINLY BEEN the originator of encouraging personal growth for every team member as a core value of the company, but everyone at TCG understood and appreciated the importance of creating a professional environment that put a premium on the personal lives of all of the consultants. For that reason, TCG was the philosophical antithesis of so many of the other consulting groups that put billable hours above job satisfaction or put unreasonable pressure on their people to ratchet up those insane (and soul-killing) hours that too many in the industry insist on.

Still, everyone at TCG was equally respectful of the need to get the work done, and late evenings sitting around the conference table discussing a client project were not completely unheard of.

The Carter Supply implementation was one of those projects that required a little extra time, and so, the fact that most everyone else they knew was already sitting in traffic on their way home was nothing but just another day at TCG.

The TCG team Steve and Hannah had gathered together included Denise, Alexa, CJ, and Rian. Denise's background as a software engineer helped her lead the programming side of implementation.

She was backed up by Alexa and CJ, who had interviewed the Carter Supply teams and mapped out functionality protocols within and across departments. Alexa had served in the military and was a natural-born people person, while CJ had crossed over to TCG after half a career at another firm and possessed an uncanny handle on all things process. Rian, one of the newest members of TCG, had spent the previous couple of years as a cellist in the city's orchestra. She was hypersmart and hardworking. Steve hoped that by learning from these TCG veterans, she'd become one of the company's best leaders.

From the implementation's outset, the four functioned as a well-oiled machine. It wasn't by accident that Steve, Maria, and Hannah had matched them together for this project, one of TCG's toughest of the year.

"All right, let's get back down to it, then," Steve said. "The folks at Carter Supply haven't offered us any specific examples yet, but they clearly have some discomfort with making the transition, and I think this uneasiness is leading them to conclude that any problems with the implementation are caused by something we did."

"None of that's exactly unexpected," Alexa said. "The systems they had in place and were using day to day were really outdated. I have to believe that the transition from their existing processes to the functionality that our implementation gave them would have been something like switching from a rusty old Schwinn bicycle to a Harley Fatboy—maybe a little challenging at first, but once you get the hang of it ..."

"I think that's right," Denise added. "We just went live a couple days ago. The hiccups sound normal. Of course, we know that when a company goes through an implementation like this, it's understandable they feel a little vulnerable. I mean, we've been through this a hundred times, but it's their first experience with the process, and they can feel like maybe their business is being interrupted."

"The Valley of Despair," CJ concluded.

"I'm very confident that's ultimately the case," Steve said. "Still, I wanted to get everyone together to triple-check the current status as we see it."

"One thing that was certainly unusual," CJ said, "or at least distinguished the project, was that it was a rescue from another group's implementation."

Steve knew this was a factor that had to be considered. "No doubt. Not all of our projects originate with us, and while taking over a project midway through the process does bring along its own unique challenges, it's nothing that we haven't handled before. I'm confident in our ability to minimize the stress involved with taking over a project that's maybe drifted off track and putting it back on track toward a satisfactory completion."

Alexa nodded. "With all of these new guys who are just flooding the market with promises and bravado but who lack the experience, expertise, or commitment to back it up, it's nice to be on the team that never overpromises, has the technical know-how to back up what we say, and always follows through on a project."

"There's something else," Steve said. "Y'all have to keep in mind that Carter Supply came to us as a referral from Janelle Squire at the Lovejoy Corp. More and more of our clients come to us on the strength of recommendations from clients for whom we've already performed these operations. That's important to me."

"It's important to all of us," Rian said.

"All right, let's start at the beginning," Steve said, nodding to the team. "Once we'd done the deep dive into their business and performed our analysis, what was the next step we took?"

"We sat down to discuss Carter Supply's roles and responsibilities in the implementation. Ben Mixon came on as their project sponsor and assumed the responsibilities of making sure our deliverables met with their corporate strategies. He was our go-to gatekeeper," Hannah said.

"Right," Steve agreed.

"Then Craig and Walker were responsible for quality assurance and leading the steering committee, providing overall direction to our collective team and managing the client-side implementation activities," said Alexa.

"And, of course," added Hannah, "Brittany Brower took the helm for change management. She coordinated the transition with the employees of Carter Supply as a whole. She delivered the training strategy, but now I'm thinking her team struggled to fully embrace the training, since this was the second time they had to learn a new system within a few months. If you ask me, Brittany faced a team for whom a new system had lost some of its charm."

"I think that's right," Steve added. "There were certainly challenges in that regard here. I don't think it's unexpected when you have a company like Carter Supply that's become so used to doing things 'their way' they may underestimate the degree of training their own people will require, especially the second time around."

"And leadership seemed to confirm this when we spoke to them today, although we didn't speak with Brittany," Hannah noted, catching the others up on their day's visit.

"All right, after leadership roles were assigned, what happened next?" Steve asked.

Everyone in the room knew this one, but Hannah was the one who offered the answer. "We walked them through Phase Zero to quickly get to the areas that were most critical to the business and further ensure the success of the implementation."

"You know," Denise interjected, "I realize I'm preaching to the choir here, but I still marvel that we're the only consulting firm out there who offers a program like Phase Zero. I mean, none of our competitors offer this kind of service, where they sit down with the client for several days in order to review every single step of the processes that matter. No way we could do that with the entire business, but there's not much point in reviewing business basics that all companies do, such as cutting checks or creating journal entries. I'm not saying that it's not an awesome program that provides for an infinitely better experience for the client and a better implementation overall—because it absolutely is—I'm just saying that maybe we could save the client some money."

> That makes our unique process, including Phase Zero, infinitely more "cost effective" for our clients, and it allows us to develop the reputation we want as a consulting firm that distinguishes itself by always putting clients' needs first.

"I hear you, Denise," Steve replied. "Lots of people think in terms of cost-effectiveness, but TCG is focused on providing real value to our clients. That may seem like a small distinction, but it's absolutely critical, and the short-term cost is Phase Zero. In the long run, that makes our unique process, including Phase Zero, infinitely more 'cost effective' for our clients, and it allows us to develop the reputation we want as a consulting firm that distinguishes itself by always putting clients' needs first."

Denise nodded. "I see that. The more we put into Phase Zero, the better the long-term value."

"Exactly. Hannah, you were saying?" Steve prompted.

"So, on the first day of Phase Zero, we sat down with them and reviewed our analysis of the business, including potential gaps. Alexa then went over the risks we identified along with associated mitigation strategies so that everyone was aware of the reasons behind the buffer we had to incorporate into the implementation schedule," she said.

"Next, we reviewed the plan that we had drawn up to address their specific requirements and documented recommended solutions or process redesigns for the identified gaps. We went over everything so that they could see exactly how we modified their approach to better reflect their previously defined metrics and overall goals of the company.

"Then Denise and CJ went over all of the technical aspects of the implementation. We defined all of their third-party integrations and then itemized and prioritized them so that there was a set plan in place for the actual implementation."

"And were there any problems with these steps?" Steve wondered aloud, because for all of the back-and-forth they had done that evening, they still hadn't been able to determine what exactly had gone so wrong.

"No," Hannah answered. "It's like I told you: they signed off on every deliverable. As always, we are diligent about getting signatures not only for each step of Phase Zero but for subsequent phases as well. This is how we confirm that the client not only has a complete understanding of what has been delivered but also is committed to following through with next steps. There's something about putting pen to paper that makes people take it more seriously. It's not necessarily a fun conversation but always a critical one to have."

"So, who handled the third step of our Phase Zero?" Steve asked. "CJ, that was you?"

CJ nodded. "Yes, that was me. I drafted a process flow for the overall implementation schedule and reviewed it with the steering committee so that everyone in the company would know exactly what was happening and, more importantly, *when*. Obviously, this was a little different than normal for a number of reasons. First, because we inherited the job from that other consulting firm, we had to do some remedial work to make sure we agreed on a baseline. It wasn't like we could just go in there and do what needed to be done right out of the gate. We had to clean up the other firm's mistakes. Second, because of those activities as well as a goal to be live by the end of the year, we had a rushed time schedule on the project," CJ continued. "So, we really had to find ways to accelerate everything."

"Could that expedited time schedule have caused any problems with the implementation?" Steve asked.

"That was our initial concern, since any decision or functional delay would eat into the small amount of buffer we had, but I really don't think that's the issue," CJ said. "We sped up the time schedule because the client had certain demands, but you know we wouldn't put anything into motion unless we were absolutely certain that we could deliver on it one hundred percent."

"No, I know that's true," Steve said. "So, what about the last part of Phase Zero?" Steve asked the group.

Denise said, "I went through all of their legacy data with them so that we knew exactly what needed to be converted."

"And what was the level of their participation?"

"We ran into the typical issues of duplicate client and vendor information as well as data that needed to be scrubbed further, but we got through it okay," Alexa said. "They wanted to be responsible for conversions to save a little money, but they were kinda slow in getting it done, so we ended up helping more than expected. It was a big task for the CIO and the CTO, since they were the only ones who truly knew the data and what needed to be converted."

"Walker and Craig." Hannah filled in those blanks.

"Right," Denise agreed. "I think they seemed hesitant to let us help, but we got everything done in time. If there's a problem, I don't think it's going to be tied to data conversion."

Steve was struck with a thought. "Hey, gang—as I recall, there were some last-minute changes to the business that we had to incorporate into the plan. Right?"

Alexa sat up. "If you're referring to the acquisition of Betsy's Produce and Restaurant Supply, you would be spot on, boss."

Steve shook his head. "That's it. Sprocket manufacturing and restaurant supply. Talk about diversification."

Alexa nodded. She thought the story was too good not to tell again. "Mr. Mack has family who have married into a small produce and restaurant supply company. But instead of continuing to run that produce supply as a separate business, they brought it in under the Carter Supply corporate umbrella, and that meant we had to incorporate it into our system during the implementation."

"Well, that's the beauty of our system, right?" Steve commented. "We can blend even the most seemingly unrelated companies with drastically different requirements and processes into a unified organization on one business management platform."

"That's exactly right," Alexa said. "But we not only had to design a system that would track receiving raw materials, manufacturing them into a quality product, and delivering premium sprockets to a fairly large list of corporate clients—we also had to integrate all of that with a business that's buying a hundred pounds of tomatoes and selling them—often for cash—to local restaurants around town."

"So, were there any problems with that?" Steve asked. "I mean, obviously there are some challenges merging systems for a large-scale manufacturing and distribution organization with what's essentially a small, local, cash-based business."

Alexa shook off the suggestion. "There were some challenges for sure, but as you know, the system can be configured by the right people to integrate just about any business, including sprockets and produce. And fortunately for all concerned, we have the project team that can work those kinds of miracles."

"You're right," Steve said. "We've all worked on projects that have incorporated the most seemingly unconnected businesses into a single system. That's the beauty of this ERP system and the hallmark of how we're able to configure it on our clients' behalf and make managing any number of different businesses as coordinated and functional as managing one."

"Just one more thing," Alexa said.

"What's that?"

"I think the previous firm messed up so badly, alienating the Carter Supply team, that it became difficult for them to see us in a spirit of partnership."

"We've all worked hard to make TCG one of the preeminent consultancies in the country for overseeing the installation and implementation of our cloud-based ERP," said Steve, "but the end result absolutely reflects the degree of partnership we're able to form with our clients. There's no doubt that when the client comes to us excited and enthusiastic about the prospects of what we're going to be able to achieve together, it makes for a much more successful implementation than those rare occurrences when someone's resistant to the process and refuses to change."

"Well," Hannah said, "every implementation can be a good implementation, and we'll figure out the bugs in this one in no time."

"That's right," Steve said. "None of those shoulda, woulda, couldas change a thing for us. We will make sure that the client is satisfied with the work we've done by proving the system is functioning as expected. Listen, I want to thank all of you for staying a little late to go over all of this. You're my closer group."

Denise smiled. "LOL—is that the company's name? The Closer Group?"

The team shared a good laugh. They were all proud to go the extra mile. It bonded them together.

"On a related note," Steve continued, "I'm so looking forward to our company trip."

The team smiled.

"Where are we going this year?" Denise asked.

"Cancun?" Alexa guessed.

"Bermuda?" Denise wondered aloud.

Steve turned to Hannah. "Have you sung karaoke in Bermuda yet?"

"Not yet," she said with a smile.

"Bermuda it is."

TVGENIUS

Are Your Expectations On Track?

If there's one misconception I could dispel about the ERP implementation process, I think it might be the misguided idea that it is a process we undertake and carry out by ourselves in a vacuum.

That's not the case.

Rather, I see our role as one of collaboration—a very active collaboration—but it is absolutely a partnership in collaboration.

The fact of the matter is that implementing the system requires active participation from our clients.

We design, perform, and oversee, but in the most fundamental way it's your implementation.

We absolutely hold your hand and walk you through the process, but you need to own the process—and the change that occurs as a result.

You need to be vocal and responsive to us throughout the project.

And, maybe most important of all, you have to take an aggressive role in accepting the change, acting to mitigate its effects, and empowering your employees to adapt as effectively as possible. Simply making sure that your people are thoroughly briefed and well trained on new protocols and processes can make all the difference between a relatively painless and completely successful implementation and, well, something that might be the subject of a riveting mystery novel...

One of the principal components of the preparedness mindset at The Vested Group is our unique Phase Zero stage of your project.

In the preplanning Phase Zero stage, the old adage that an ounce of prevention is worth a pound of cure *really rings true. Phase Zero is where we finalize and publish the project schedule so that everyone knows what's supposed to be happening on the project and when. Down the road, strict adherence to that project schedule will not only keep our people on pace but will serve as a mechanism to make sure that our client partners are holding up their end of the bargain and fulfilling their various responsibilities. We review our solution design so that the client knows exactly what we have planned for the implementation. And we get everything in writing so that the client has the clearest possible idea of what is happening with the implementation and their business.*

NEXT STEPS:

While it's not possible to see around every corner, our planning and kickoff methodologies are solid enough to allow us to avoid most of the hiccups and surprises that can throw off a project. If you'd like to take a look at how this works in the real world, you can review the "Scope" section of a sample SOW (statement of work) from The Vested Group at joelpatterson.com/the-book.

CHAPTER 10:

JEFF AND THE SANDBOX

THE NEXT MORNING, there was a big box of donuts from Rainbow Donuts on the conference table. Steve hadn't even asked Maria to get these for the team who stayed late last night and came in early—she just knew how to take care of people. He popped his head in as he walked by.

"Thanks, team," said Steve to Denise, Alexa, CJ, and Rian, who'd reassembled around the donuts. "Now I have to get to the bottom of what Jeff knows."

Although he'd had early success with his own consulting group, Steve had always found something missing in the enterprise.

And so, when he set about to create TCG—a new and improved version of his initial foray into private consulting—his mind was very focused on making up for what he saw as the few but significant shortcomings of that previous endeavor.

At the very top of that list of "things to do differently" was a conscious decision to step away from the "one-man band" situation and purposely look to incorporate others into the creation of the business.

Steve was more than confident that he had the technical expertise and business acumen necessary to make a success of his consulting group (he'd already enjoyed a successful career in the "big leagues" and done it once before on his own), but he knew that in order to really achieve the sort of internal drive for excellence required to offer truly outstanding service to his clients, he would need to incorporate the visions and perspectives of others.

For that reason, Steve had intentionally sought out a partner to make building TCG not just another solo show but rather a partnership—and he'd hoped it'd be a much richer experience because of that camaraderie.

He didn't have to look far for his top candidate.

While Steve's professional experience had been more or less confined to technology consulting in one form or another, Jeff had a solid background with customer service in the take-no-prisoners world of commercial freight delivery for a national corporation.

That professional background would have been all he needed to fill the role that Steve had envisioned for him at TCG, but Jeff also had another very important credential: Jeff was Steve's brother-in-law.

That wasn't to suggest that there was any nepotism at work—just the opposite.

What Steve was looking for in order to build the team of his dreams was a management foundation that functioned as closely as family—so what could possibly be a better fit for that position than actual family?

There was still one more attribute that Jeff brought with him. Not only did Jeff have that strong professional background and the bond of brother(-in-law)hood with Steve and to TCG, but he had something even more important than either of those two strong credentials. Jeff also brought to the team a temperament that was an invaluable (maybe even necessary) balance to Steve's own personality.

Whereas Steve was focused on modern concepts of collective corporate culture and was committed to making the work environment as pleasurable as possible in order to build a sense of team for the group, Jeff had a different ideal of what "being a team" meant. Jeff's character and outlook on life (and business) had been forged on the hallowed fields of Texas football.

From his earliest days in peewee football to his time in collegiate ball, Jeff had learned dedication and discipline not only as an aspect of the game that he loved but also as a way of life and of conducting oneself even off the field.

Two-a-days in August.

Leave it all on the field.

Know your role and execute, execute, execute.

The great tradition of Texas football had also instilled in Jeff a sense of competition that did not make concessions to the word *quit* and would not acknowledge the word *lose*. This wasn't always apparent in his interactions with others in a business setting—although he was seriously not someone to challenge to a "friendly" game of Ping-Pong.

Or pool.

Or just about anything.

Of course, Jeff's sense of competition came out in the workplace most intensely when he was measuring his own performance and raising his personal bar to demand more of himself than he had exhibited the day before.

This was his personal standard, but it was also the measure that he applied to everyone else in the office—a metric that all of the team members earnestly wanted to satisfy every day.

As a result, he introduced a stricter, more regimented presence to the office.

It was this "game day" intensity that Jeff brought to the office every day that balanced out Steve's earnest desire to discover what more he could do for his people.

Yin and yang.

A perfect balance.

But what worked best about Jeff's influence on the office was that he held absolutely everyone to the same standards. Even Steve.

He was prepared to do anything to rescue the Carter Supply project, which in his mind was bogging down in the fourth quarter of a championship game.

"What do they want us to do?" Jeff asked Steve, who had just stepped into his office and brought up Carter Supply.

"We haven't quite gotten to that part just yet," Steve said.

"Well, that's what we need to find out," Jeff said. "I was all sleeves-rolled-up-and-in-up-to-my-elbows during the technical background review of this implementation, and I'm confident that we nailed the agreed-upon scope with the functional and technical design docs and subsequent development efforts."

"I'm sure it's something that we can work out," Steve started. "We still haven't gotten their punch list identifying just what they think has gone wrong, so it's premature to come to conclusions, but it's likely the Valley of Despair. I mean, you guys didn't have any problems outside of the typical go-live issues, right?"

"None," he said. Then suddenly something clicked in his head, and Jeff's eyes flashed with a glimmer of recognition. "Except …"

"Except what?"

"It wasn't anything that I took super seriously," Jeff prefaced. "But while we were working on the implementation, we were getting some screwy results on tests run in sandbox."

"In sandbox?" Steve asked quizzically. "That's just the parallel testing environment that we set up to give us a platform on which we can test that everything is working the way it's supposed to before going into the system. I mean, in case you don't know what the purpose of a sandbox is …"

"Yeah, I might know a thing or two about it!" Jeff said. "For the most part, everything tested fine, but while we were working, there would be results in sandbox from time to time that suggested maybe everything wasn't as perfect as we needed it to be. But then we'd go back to run more tests and everything would be perfectly fine, and we couldn't reproduce those results in the production environment."

"I don't understand," Steve said. "You're saying that you were getting screwy results in sandbox and that they'd correct themselves in production?"

"Well, that's just it," Jeff said simply enough. "One day we would go and look at the results from the test environment and they'd be off. No big deal, right? We'd just have to go back and tweak a little this or that, but then we'd go back to take another look, and all of a sudden all the problems that we'd identified would have resolved themselves completely."

"That is screwy," Steve admitted.

"It worked both ways. Things appeared to go off track on their own, then work themselves out, and then go all screwy again. Back and forth, back and forth."

"Did you ever get any possible explanation?" Steve wanted to know.

Jeff shook his head. "No. I'd never seen anything like it. We'd have something totally locked down, then come back to take a look at it, and it'd be off. Then we'd come back to take another look at it in order to evaluate just what had to be done, and it'd be correct again.

It only happened a handful of times, so I didn't think much of it. But it does seem highly unusual."

Steve was puzzled, as well. "That's bizarre. Any theories on what it could have been?"

Jeff shook his head. "As soon as Carter Supply broke for the holidays and it was only us there for the implementation, the situation seemed to resolve itself, this time for good."

"That's odd." Steve wondered to himself about the possible causes and cures. "I'd like to take a look at those test scenarios and results when you can get them together. Maybe there's something in there that would give us a clue."

Jeff nodded. "I'll go back into sandbox and get them for you as soon as I can. Let me know if you want me to go over them with you or if there's anything I can do to help. Anything at all."

Steve nodded. But in that moment, he was grateful that he'd had a hand in bringing together a corporate culture in which "tell me what I can do to help" meant just that.

Steve got up from his chair, took a couple of steps, and then turned slowly, as if he'd just remembered something important. "There is one more thing."

"What's that?"

"Besides sandbox, was there anything else that went wrong, that you think is worth looking at as we try to get all of this worked out?"

Jeff thought for a moment but shook his head. At first.

"I will tell you," Jeff finally said. "Carter Supply, well … I don't think they realized the depth of the change we'd bring. It's funny, because we design these software systems to run exclusively in the cloud, and that description alone would make people think that the functions do what they do—or *don't*—independently, without any human input. But what people don't realize is that the systems are designed for humans to use, and their effectiveness is absolutely dependent upon the humans that are working with them."

Steve nodded. "Absolutely. No argument here."

"Right. So even though we could implement the exact same system in ten different companies—"

"Well, we wouldn't do that," Steve objected.

"No, of course not. Not in real life. Obviously, we tailor everything for each specific company, but I'm saying hypothetically that if we did put the same system into ten different companies, you'd get a spectrum of results back, based not on the system—which would be identical—but on the people who were using it."

"Absolutely," Steve said.

Jeff shrugged. "Well, it takes a one hundred percent commitment. And I think anytime there are complaints about 'problems' onsite without anyone being able to identify just what those problems are, you have to look first at the possibility that there could be some sort of human error at the user level."

"And you think user error may be the biggest problem that they're having with the system?"

"I think so. With an interface that depends on interactions with different departments and how everyone uses it, if a few aren't a hundred percent up to speed, the system won't function as well as it should. It's like agreeing to be a contestant on *The Bachelor*. You've got to be there for the right reasons—to fall in love. If you're not there for the right reasons, then there's no love to be found and you go home."

"Jeff?"

"Yes?"

"That's an interesting suggestion. And I'm not here to judge, but I think there's a strong possibility that maybe you watch way too much TV and haven't quite figured out the farce of reality shows."

"Happy to help. That's how we do things at Trust and Credibility, Guaranteed."

"I've told you from day one—that's not what TCG stands for."

"Don't worry." Jeff winked. "I'm using it all the time now. It'll catch on."

TVGENIUS
How Do You Know It Works If You Can't Test it?

The Vested Group utilizes a test environment protocol commonly known in our industry as a sandbox (as in, go play in the sandbox). Essentially, it's like the Mirror Dimension from Doctor Strange or a parallel system if you're not into the whole Marvel movie thing. Anyway, by creating a parallel system, we can conduct test runs of all the stages of the implementation to make sure that there aren't any problems or negative interactions with the newly defined process and/or third-party systems with which we may integrate. While having a test environment is a fairly standard practice, knowing how to properly track and manage all of the changes, updates, refreshes, and the like and keep a good grasp on the state of the test environment versus the state of the production environment is not for the inexperienced.

NEXT STEPS:

Validate that your project schedule has an adequate testing phase. Not sure where to start? Please visit joelpatterson.com/the-book to view a UAT (user acceptance thingy) guide to get a good feel for where you stand.

CHAPTER 11:

THE NEW OFFICE

THERE WAS NO DOUBT that the original TCG offices in downtown Plano were both state of the art and comfy cozy.

Their 15th Street location had been constructed in the nineteenth century as a feed-and-grain store, back in the day when those types of enterprises were the country equivalents of a Walmart or Costco.

For that part, many people had written off the entire downtown district of Plano as a fading chapter of the area's history and one that was better left in those bygone memories of a different era—especially in a North Texas real estate market that was filled to overflowing with shiny new office buildings that offered tenants the most fashionable accommodations any business could want.

Still, Steve had not only seen the potential in the tired old building; he'd seen something much, much more.

When he looked at the building, he saw through its rough condition and the largely forgotten downtown neighborhood it occupied.

No, from the very beginning, Steve had looked through all of that and seen only the future. He saw only potential and possibilities, not only for his own company but also for the neighborhood as a whole.

It was a community that he was certain TCG could become engaged in as an active member.

TCG had started as a tight band, a handful of individuals united by a vision, and those original offices were more than sufficient to offer all of the space they could need to spread out and get down to work.

And socialize. And generally have a great time at work. And not view work as a "four-letter word."

The problem, however—and Steve had always realized that it was a very good problem to have—was that their unique customer-focused services had proven successful in the marketplace.

So successful, in fact, that all of the good word of mouth and direct recommendations from their clients had attracted still more clients.

And then more clients still.

Until finally the change that Steve had wanted to avoid could not be contained any longer, and physically expanding the office space became a necessity just to handle all of those new clients.

And so, they had set about hiring a few new consultants.

Not too many at first.

Steve was worried that if TCG grew too fast and hired too many associates, all of those new recruits would dilute the personal closeness and chemistry that he knew was an essential component of the group as a whole.

The problem here—and, again, it was a good problem to have—was that the new hires had been awesome.

And they *really* were the best in the business.

So, not unexpectedly, the work that they had done was top notch, too.

And because of that, there had been even more recommendations, and that all led to even more clients knocking on that antique-chic front door.

And on and on, until before anyone realized what had happened, there were more consultants than there were workstations.

That meant that every morning Maria was like an air traffic controller for team members, stacking them up like jets incoming to DFW. "Take Cindy's workstation while she's onsite at a client today," or "Why don't you squeeze in over there at Casey's desk for now and we'll figure something out."

The "rearrangement arrangement" worked for a while, but every new hire only made the office congestion problem worse.

No matter how they tried to get by with sharing desks and working the conference rooms or breakrooms, eventually the team reached a collective conclusion that there was really only one solution to their dilemma.

No, they couldn't sell the place.

Everyone at TCG had a real emotional attachment to their storefront location, and they all agreed that they simply couldn't stand the thought of ever letting it go.

Instead, they all agreed that the best option would be to simply expand their operations.

Of course, that led to a very difficult question: *Where?*

There was reason to consider downtown Dallas.

Or farther out into some of the technology-centric neighborhoods throughout the metroplex.

There were nothing but good options; still, like so many other things in Steve's life, the perfect answer to that troubling question just sort of fell into his lap.

The property was a corner lot just a five-minute walk (or two-minute golf cart ride!) from their current offices and offered more-than-ample parking. This was admittedly something that was occasionally an issue in their storefront location.

And if all of that wasn't enough—and it was!—an added bonus

was that the place was also situated across the street from the DART tracks, which meant easy commuting to almost anywhere in the metroplex, including downtown Dallas and DFW airport.

There was no doubt about it: the new location was a true gem of a find in a real estate market that had exploded and then continued to grow as Dallas's near-daily expansion pushed the boundaries of the metroplex farther and farther.

So, although he was by nature a somewhat careful and methodical man, Steve didn't hesitate for more than a moment or two before the deal was made, the papers were signed, and the property was theirs.

Steve had spent so much time trying to find solutions to the problems that didn't exist on the Carter Supply implementation that he hadn't had the time to stop by the new TCG offices in a while.

He'd missed seeing the place, and when he laid eyes on their new offices early the next morning, he was relieved to feel a certain rush of excitement that he was quite sure he'd never lose.

The place was just mortar and bricks, of course. He knew this.

Their new offices might have been just another building, but Steve saw the place as a testament to his dream, a walk-in-the-front-door monument to the realization of his goal of creating something that was bigger than himself, something more than a consulting group—a living philosophy of a better way to serve clients and do business.

He loved the way the new building reflected all of that.

In what was to become their conference room once all of the construction was over (whenever that was), Steve found his wife, Maria, waiting for him with a beaming smile. At their feet was the biggest single piece of wood that he'd ever seen in his life.

"What's this?" he asked.

"This is a plank cut from the legendary Tarrant County Court-house oak tree," Maria announced proudly. "At one time it was the tallest oak in all of Texas. It's going to be the main conference table," Maria announced.

The plank of wood was forty feet long if it was an inch, eight feet wide, and six inches thick. It was an enormous hunk of wood, but it wasn't anything more than that.

"And just how are we going to turn the expensive conference-table-that-isn't-a-conference-table into a conference table?" Steve asked.

"I'm going to do it," a voice answered from behind.

When Steve had set about on construction of TCG's expansion offices, he knew that the most important thing about the project was getting the right general contractor to oversee all of the work. So, he made sure to go out and get the very best that he knew: his dad.

"Dad," Steve said, "I appreciate everything that you're doing here with construction, but turning this"—he gestured at the wooden slab—"is more than just a construction job. It's more than just furniture making. It would require—"

"A work of art," his dad said matter-of-factly.

It wasn't that Steve doubted his father's capabilities, but he wasn't certain that he could pull off a project of that size and nature either; at least, not without turning the once-tallest oak in Texas into an insanely expensive end table—or toothpick.

"Look, Dad, you know no one thinks more of you than I do, but—"

"My entire life," his father said. For a moment the comment seemed like a non sequitur, until he followed up with his explanation. "For my entire life I've been working with wood. Beams here and rafters there. Just construction materials. But this …" Steve's father gestured to the slab, and his eyes widened with appreciation. "This is something so much more than that."

He crouched down next to the oversized plank and rubbed his hand across it like it was a living thing and he was showing it some affection. "Look at the colors, the patterns of the grain, the number of rings. This isn't just a piece of wood; it's a piece of history. Think of all of the things that this tree must have seen: The Dust Bowl. Apaches and Comanches. The Civil War. Even the start of the Chisholm Trail for longhorn cattle drives."

He turned to his son. "And I'd like the opportunity to turn that piece of history into something that you're going to build history on. Just like you've built this company, are building this company, I'd like to do something for you that's grounded in the past but designed for the future."

With that, the fate of the plank was sealed.

"And your mother, it must be said," he added, gesturing to the bare light fixtures, curtainless windows, and general lack of décor, "has one hell of a plan to furnish the rest of this. She's the most talented interior decorator you could ask for."

"He's right," said Maria. "I've seen her mock-up. Incredible. Wait till you see the bathroom. But we want it all to be a surprise for the grand opening."

Steve was so thankful, he didn't even know what to say.

With TCG, Steve wasn't just building a company; he was putting down the roots for a legacy that he hoped would long survive him and Jeff.

In that same way, Steve understood that their clients weren't simply entrusting them with their businesses but also with their life's work, with their own legacies.

And if he could get sentimental about a tree turned into a table or a well-furnished bathroom, then Steve thought there was nothing wrong with finding a similar emotional incentive within that relationship between TCG and its clients: a certain sort of trust, but something more, too.

They weren't simply implementing software systems for these businesses; they were helping their clients to fully consider their futures and to realize their dreams, both in securing the near future and in creating a legacy. They were giving them a stronger foundation from which they could plan their futures and build something that would last beyond them and be suitable to pass along to generations following along—like a beautiful table.

> They were helping their clients to fully consider their futures and to realize their dreams, both in securing the near future and in creating a legacy.

Steve smiled like the others. "That'd be nice, Dad. And do you think you could drill some holes here in the center?"

"Drill holes? What for?"

"So we could put in some extensions. You know, for the laptops. Maybe run the wires for a speakerphone."

His wife and dad looked at him as if he had proposed scribbling a Sharpie mustache on the Mona Lisa, but no one said a thing.

"Or we could just leave it as a solid plank," Steve offered.

The three of them smiled.

Decision made.

TVGENIUS

What Does It Mean to "Give Them Your Shirt"?

One of our company core values is "Give Them Your Shirt." While we obviously didn't coin this phrase, that doesn't change the fact that there are significant reasons this made it onto our core-values list.

Sure, we do the typical things you would expect from a "Give Them Your Shirt" company—we've volunteered with charitable area organizations over the years and contributed to fundraisers, drives, and the like as they make their way onto our radar. We marathon and 5K and bike our way across the metroplex for good causes. We support our local community as a conscientious investor.

But what really makes us choose "Give Them Your Shirt" as one of our core values is the smaller everyday examples that happen all the time around here. When we are making hiring decisions, one thing we always consider is something we refer to as the Sunday Test: If this person was in a bind and needed me to come to the office on a Sunday and help them with a project or task, would I lend a hand, and, more importantly, would I feel good about it? *If the answer is yes, let's assume this candidate would do the same for us and has values that are aligned. This tells us they will mesh well with our culture and bring their own TVGenius "flair." That mutual respect and desire to* give your shirt carries over into every project we do, both externally and internally.

NEXT STEPS:

Sometimes *giving your shirt* is easier said than done, particularly on a company-wide level. To help navigate this path, we partner with an organization called Conscious Capitalism (www.consciouscapitalism. org). They offer guidance for business leaders who want to elevate humanity through the idea that their business can have a purpose beyond profit.

CHAPTER 12:

BRITTANY

STEVE AND THE OTHERS had intentionally designed TCG as a relatively relaxed work environment, but that didn't mean there wasn't a very firm structure existing underneath.

There were certain rules and policies, practices and procedures, all of which were followed by everyone—from the original partners to the newest of hires—without exception.

"Jeff," Hannah said from her workstation.

He was seated across from her and looked up. "What is it?"

"I've got Brittany Brower from Carter Supply on the line. Could you talk to her for me?"

One of the policies that was followed strictly at TCG was simply that everyone handled their own projects without unnecessarily delegating those duties or invoking any sort of positional hierarchy.

That is, if a consultant had a call to make (or take), they were expected to get it done without constantly passing any of the issues to be covered up or down the chain of report. That sort of assignment of responsibilities wasn't put in place to avoid taking on those challenges at a senior level but rather were designed to develop confidence and a sense of responsibility for projects (and the group, in general) among even the newest of hires.

That's why Jeff thought Hannah's request was an odd one. "Because?"

"Because I'm afraid that maybe she's gone beyond talking."

Although he didn't understand the full nature of Hannah's request, Jeff picked up on the cue that she needed him to step in, and so Jeff was quick to pick up the phone. "Certainly."

He touched the button, and the line went live. "Brittany, how are you?"

"How am I? I'm really frustrated, to be honest. Nothing's working like we told everybody it would."

Brittany Brower, the VP at Carter Supply and helm of the change committee, had had her hands full from the beginning. Her responsibilities included ensuring that all of the users on the Carter Supply side were all sufficiently trained and acclimated to the new processes and protocols that were necessary to interact with the new system during the course of their own workdays.

Jeff understood that given the sort of finger-pointing that might well have taken place in the corporate halls of Carter Supply since go-live and the resulting difficulties, this situation might possibly have put Brittany square on the receiving end of a blame game. He understood that this had to be a terribly difficult place to be.

So, Jeff didn't take anything away from her frustrated tone other than that she was understandably upset.

In the same way that Jeff had the perfect demeanor for managing his partners through his unique balance, he possessed an equal charm in handling clients, including the ability to talk to even the most agitated person in a way that restored calm to whatever difficulties had arisen in their communications.

"Help me understand what's happening, Brittany," Jeff said.

"Well, it's not working the way that it should," she replied.

"Brittany," Jeff said, with authentic curiosity, "my team is on this, and it's already their top priority. But while I have you on the line, I did have a question. You helped us lead six training classes, one for each of the company's departments. You also joined us in biweekly training to become one of a handful of superusers so you could answer any questions that may arise from the Carter Supply team. What's your thought today on the depth of training? Do you and your team of superusers feel prepared to handle all the questions coming your way?"

Brittany breathed a sigh of relief. "To be honest, if you asked me a week ago, I would've said we were prepared. I even said the same to Hannah and your team at the time. But ever since the go-live happened, my team and I have been overwhelmed.

"What changed?"

"I guess we worked so hard to prepare for the first firm's implementation that we didn't spend the same amount of time training on this one. Even though it was different and required a new set of training, we thought we could apply what we'd learned to this one."

"That sounds pretty reasonable to me," said Jeff. "I would've done the same thing. Nobody likes thinking any effort is wasted. Was there anything else that seemed to go wrong in the training?"

Brittany paused. "Remember how sandbox didn't function properly during every training?"

"Yes, I was just thinking about that, too—how odd it was that the sandbox was so glitchy," Jeff said, catching Hannah's eye.

"Well, I didn't make the connection back then, but the glitches always seemed to happen right before a training session. It made it really hard for me to look like I knew what I was doing!"

"A glitch right before training? That does sound hard."

"Jeff, thank you for listening. Hopefully some of what I'm saying makes sense and can be helpful to you as you figure this out. I agree we'll need more training, and if that's what it takes, I'll rally my superusers."

"Just what I was thinking," said Jeff. "Denise and the gang can meet with them tomorrow and help answer all the superuser questions. They should be able to handle the tasks from there. You can rest easy—we'll take care of this in a jiffy. That's why we're called 'Trust and Credibility, Guaranteed.'"

"Is that really what TCG stands for?"

"It sure is."

TVGENIUS
Do You Appreciate That Failure Leads to Learning?

We all try.

Still, sometimes there are situations in which people's better instincts succumb to the pressures that we all face. When that happens, occasionally communications can become difficult. You know what I mean …

When that happens, I believe personally—and have infused this outlook into our culture here at The Vested Group—that I'm all for giving those folks the benefit of the doubt.

In servicing our clients, I have always contended that it's essential that everyone remembers we are all on the same team, with the same goals in mind—and that any degree of division only frustrates the realization of those goals.

And just as one of those making my way through this crazy thing we call life, I have found that it is always important to keep in the back of your mind that everyone is dealing with things outside your view of the world. Often even the most personal of vitriol isn't personal at all but rather a misguided action resulting from too much internal pressure. My friend Bradley Callow often liked to remind me: "Go easy—we're all human after all."

Whatever the cause, it's always best to respond calmly and with the utmost kindness, humanity, and compassion, exactly as Jeff did in this situation.

NEXT STEPS:

One of our least talked about but most important core values at The Vested Group is "Appreciate That Failure Results in Learning." Nobody enjoys making mistakes, but you should never underestimate the value in acknowledging when something is not a success and in having the humility to learn and move forward. Another spin on this is to consider that if you never fail, you aren't really trying. My son Dean and I go on a father-son ski trip each year, and while I wouldn't define falling down the mountain as a goal, we certainly give each other a hard time at the end of the day if the other didn't have a face full of snow at some point.

CHAPTER 13:

HILLARY

"STEVE, DO YOU HAVE A MINUTE?"

If Maria was the matriarch of TCG—and she *was!*—then Hillary was the chief steward.

As the head of TCG's HR department, Hillary was entrusted with what might have been the most important job of all: recruiting the best possible candidates and keeping together the one-in-a-million team that was the vital secret to the consulting group's success.

"What is it?" he asked.

"We're growing so fast," she said. "While I'm helping to put together and keep together this award-winning team that has been consistently recognized by *Forbes* magazine and a number of other authorities as one of the best places to work ..." She paused for him to insert a word of praise.

Steve just nodded. "I've read the articles. I'm very proud of what we've achieved here for our employees."

"Right," Hillary said. "Well, the more good work we do, the more our reputation for excellence gets spread around by all of those satisfied clients, and then the more new clients we attract and the more new jobs we have to do. That means we keep hiring more and more team members to staff all of those new projects. We have some more video applications to review. You know that we don't believe in résumés, because they are subject to embellishment. And even if applicants are absolutely accurate, people can have the greatest academic achievements or professional experience in the world and—nothing against them as individuals—they're just not people who would thrive in an environment as unique as TCG. At the same time, you might have someone else who, for whatever reason, maybe didn't wind up going to their first-choice school or has struggled in other situations, but when they get in a creative and supportive environment like we foster here, they just thrive and flourish and exceed even their own aspirations."

"That's what we've always tried to accomplish with our people," Steve said. "Give everyone a place—and all of the tools—that will allow them to be their very best selves." Steve smiled as he looked around the place. "And I think that we have the real-world results here to back up that philosophy and prove that it is working for our collective team and the individual members."

"So, instead of reviewing résumés, we have a policy here of just asking people to send in a two-minute video. I mean, anyone can bluff and puff their way through the single page of a résumé, but it is so much harder to fake a two-minute video. It's hard to put yourself out there even for that short a period of time and not show off your real self. We have some very unique requirements to join our team, and there's no better way to determine whether a candidate is going to fit in with the rest of us."

Steve just shook his head. "Isn't it crazy to think that there was a time when the job application process was all about your *résumé* on fancy paper and making even the simplest things sound like they were something extraordinary? Do you know what I would have given starting out to find a company like TCG that wants you to be creative and original and yourself? That thinks it's important to not only get people who are technically proficient but who can work together as a team? That truly believes that the most important thing at the end of the day is to have a relaxed family-like environment to work in?"

"Well, that time is now," Hillary said. "Not for you, but for the folks whose videos I've got cued up and waiting for you to watch with me to screen."

Twenty minutes later, Hillary and Steve had watched all of the videos that she had thought might make meaningful additions to the team.

"I like the juggler," Steve said as he stepped out of her office.

"But *not* the clown mime, right?" Hillary said as one of the many sufferers of coulrophobia.

> We'll talk to the juggler and the young consultant. I'm not sure what we'll do with the mime...

Steve paused to consider the prospect fully. "I don't know; I think anyone who can wordlessly communicate their tech experience through the ancient art of mime deserves a closer look. Bring them both in for an interview—and the young woman with all of that consulting experience—and we'll talk to all of them at length."

Steve thought better of the instructions. "At least we'll talk to the juggler and the young consultant. I'm not sure what we'll do with the mime—maybe have him walk against the wind?"

Steve did his best Marcel Marceau, but he was no mime.

Hillary cracked a smile. "Got it," she said. As she left his office, she stopped for a moment. "The Coolest Gurus?" she asked, playing the office game.

"Nope," said Steve. "Nice try."

Later that day, before everyone left for the night, Steve called Hannah aside to find out how her call went with Brittany.

"She and Jeff agreed there was a breakdown in training. She felt they'd had enough but after the go-live realized they needed more. They were overconfident, based on the training they'd started doing with the first firm. Jeff's got Denise and her squad there tomorrow to help the Carter Supply superusers."

"Good," Steve said. "Now we're getting somewhere."

"Not only that," Hannah said, "but a glitchy sandbox seemed to derail Brittany's training. Highly unusual. Have you ever heard of a glitchy sandbox? Brittany said it'd glitch the night before every training that was to happen the following day. That probably made it almost impossible for Brittany to do the training. I just can't figure out why it'd do that. Once, sure. But not before every training. It doesn't make any sense."

"Did you hear from Ben or anyone else at Carter Supply about what they think is wrong?" Steve asked.

"Just that they'll send over their report first thing in the morning."

"Great. And I'm with you," Steve said. "There's got to be more to this glitch."

TVGENIUS
How Do You Feel about Shameless Plugs?

As we point out in our definition of TVGenius in chapter 1, our people are the number-one ingredient in what makes The Vested Group special. Building an exceptional team is not something that happens accidentally—it takes a lot of specific intent. While we're certainly interested in using this book as a resource for expanding our business, our twin

94

motivator in this literary endeavor is also attracting the interest of those individuals who might be a good fit to join our team.

In looking for potential new team members, we're not only looking for the sorts of folks who possess the top-level technical skills that we demand from our consultants (that should go without saying) but also individuals who have discovered things within these pages that resonate with them and inspire them to think that perhaps The Vested Group might be the perfect environment for them.

So if you're ambitious, technically well grounded (or rounded?), and feel like our core values speak to something within you, let us know you are interested in becoming a part of the best team in the industry and contact us directly.

NEXT STEPS:

Flip through our Culture Brochure and then take a peek inside The Vested Group office for a few reasons our employees say working at The Vested Group is the best job ever: joelpatterson.com/the-book.

CHAPTER 14:

FAMILY

STEVE NICHOLSON LOVED HIS WORK. The nine-to-five (more often much later) was not a chore to him but a time spent pursuing his passion with his best friends in the world.

He loved his time with his kids: the best three kids in the world and an instant reminder of why he did what he did on those rarer-than-rare days when one problem or another caused him to momentarily ask, "Why do I do this?"

He even loved what he considered to be *his* time—the time spent alone working out or reading.

But the time he loved most of all was those moments he spent at the end of the day, when there was no one in the whole wide world except Maria and himself and they could talk about whatever was on their minds from the day just passed.

"Do you ever worry about the future?" Steve asked, apropos of nothing.

"Can Michael Scott sell reams of paper?"

"No, I mean—Michael Scott pretty much created the paper industry, of course—but I'm talking about the business. I mean, with all that we're doing with hiring new people. And the new building and all. Do you ever worry about the future?"

"Repeat: Michael Scott of Dunder Mifflin."

He didn't say anything more, and she realized his silence meant that there was something more. "What is it?"

"The magic number," he answered.

"What is the magic number?"

"The magic number. It's the number of employees that a business can have before the growth starts to deteriorate the quality of the culture. I saw someone discuss it at that conference in Denver."

She knew better than to think that it had anything to do with Denver. Or Phoenix. Or Vegas. "We're talking about your last business, aren't we?"

"No … yes. Maybe. I was so proud of the business," he said.

"It was a great business," she answered easily.

"That was such a great business," he agreed. "But it grew so fast, and before I knew it, I didn't recognize it anymore. It wasn't a reflection of me or my partners. I don't want that to happen here, not with TCG. We've all put so much into maintaining the culture, and I worry that adding more consultants, more offices might dilute that."

She nodded, because she understood. "Do you remember when we were just a young couple and we had all of the time in the world to ourselves to do whatever we wanted whenever we wanted? Remember when all of our money was our money?"

He nodded.

"And then what did we do?"

"We started our family."

"Yeah, we started our family. And everything changed, and I have to admit that I was worried about it. Worried about what the addition of a child into our life was going to do to our life as a couple."

He might have said something, but she didn't give him the chance.

"And we had Aprille. And it was different, but the change was awesome. She was awesome."

"She is pretty awesome," he said with a smile.

"We were a perfect little family of three. We weren't diminished— she made our life better. And once we had that perfect life, what did we do?"

"We had the twins."

"Yes, we had Josh and Austin. And I will tell you that I had more than a few moments when I was worried about what the new additions would do to our perfect little family. And then along came Josh and Austin. And you know what?"

"They made everything better?"

She kissed him on the nose. "Lots better. Because our family can't be dictated by a magic number or any other standard. Families don't stay together or break up because of the numbers but because of the care that goes into them.

"You haven't built a business; you've forged a family, and you put the same sort of care and concern into everything you do with them. There's no danger in magic numbers, because you care about maintaining the culture that you've worked so hard to create and cultivate. It's an extension of the family that you've built here with me, and that's why I don't worry about adding members to it, because I know you're going to make sure that they find a family here. And if our own family members are happy, they'll make our extended family members—our clients—happy, too."

CHAPTER 15:

SALES AND FINANCE

WHILE THE CARTER SUPPLY TEAM finalized the punch list and his team prepared for training and reexamined the sandbox scenarios and results, Steve thought he needed a bit more information before he could see the big picture.

And so, Steve E. Nicholson did what any good detective would do: he returned to the scene of the crime.

Steve found his way into the Carter Supply corporate offices just before the doors locked at five o'clock and found his way to the sales department.

There were dozens of cubicles, but all of them had been abandoned with the late hour.

There were, however, two guys talking by the coffee area.

Steve walked over to them.

"Excuse me."

"How can I help?" one of them asked.

"I was hoping I could talk to some people from the sales department," Steve said

"That's us. I'm Brandon," the other said. "And this is my counterpart, Ferg."

"Great! I'm Steve Nicholson. I'm with TCG—we did the system upgrade and installation of the ERP here."

Brandon raised his eyebrows. "That hasn't been going so well this week."

"We can't do much with that system," Ferg chimed in.

Steve took a deep breath. "All right. Well, let's talk about that. What exactly are the problems that you're having?"

"I'll give you an example," Ferg offered. "Whiskey and wine."

"Whiskey and wine?" Steve didn't see the connection.

"All right, so a lot of our clients are whiskey guys and wine gals, right? So during the holidays, we like to buy them some whiskey boxes and wine crates so they can give out the bottles as presents. We take care of our clients, and they help take care of their people."

Steve was surprised. "You're buying the bottles through the sales department?"

"No," Brandon corrected. "We *were* buying them. You could do that with the old system, but with this new system in place, we can't."

"Gotcha," Steve said. "The issue now makes sense. In looking over the flowcharts and our project charter, I don't recall seeing the bottle purchasing earmarked for the sales department. I think the issue goes back to when we mapped out your department requirements within the system. Is that what you meant by you're having problems?" Steve asked.

"Well, that's significant," Ferg said.

"You're absolutely right. Thankfully, it doesn't mean the system is malfunctioning. It means we didn't fully map your protocols and processes."

"Really?"

"The system we implemented is working just fine, because it's designed to ensure that proper protocols are followed by everyone.

We've just got to make sure your business processes are reflected in the system, which is causing the issue."

Brandon and Ferg nodded.

"That sounds right. What can we do about it?"

"You worked with Denise, Alexa, CJ, and Rian?"

"Sure did."

"Great! I'll make sure a couple of them come here tomorrow afternoon. In the meantime, we'll want a punch list of what protocols they should add. Is it just the whiskey and wine?"

"A punch list sounds good. We've been working on a list for our boss."

"You guys are already ahead of me. We'll get the whiskey, wine, and any other missing protocols added tomorrow. Then get you trained on how to use them."

"That's great to hear," said Brandon.

"It'll mean the world to our customers," said Ferg.

Steve had another idea. "Thank you, guys—it was really helpful to speak with you. We'll have the system working perfectly in no time. Now, I did want to see if I could help another department. Do you think there's anyone else still at the office?"

They both nodded. "Jennifer, in finance," said Brandon.

"Accounts receivable. She always closes the shop," said Ferg.

Steve left the two and passed through empty halls until he arrived at the accounts receivable office, where he found Jennifer, whom he'd befriended on previous visits.

"Burning the midnight oil, Jennifer?" Steve asked.

Jennifer glanced up from her work. "Oh, hi, Steve!" She typed a few numbers. "I had a hunch you'd be stopping by."

"Oh, did you? I'm sorry the system isn't up and running like it should."

"It's working all right," said Jennifer. "It's just got many of us confused." She nodded at the finance team's empty cubicles behind her. "I've been helping the team with questions all yesterday and today. We're still sorting out how to process invoices. It's all sorts of 'push this key' and 'go to that field.' It's not at all like it used to be. We've been doing this for seventeen years. I knew everything there was to know. If there was a problem, people said, 'Go ask Jennifer.' And now, all of a sudden, we're starting over."

"I understand," Steve said. "Now I'm realizing it was a tall order to have you be a superuser and still lead all of your department's day-to-day responsibilities. Everything's on your shoulders more so than ever."

"You're right about that. But I got faith. There were fewer questions today than yesterday. If I had to do it over again, I'd probably have Shelley be the superuser, as tech is second nature to her. She even told me, 'This system is like getting a new smartphone. It has some changes to get used to at first, but in just a couple weeks, if you take a little time to learn, you'll find you like it even better than what you had before.'"

"I love Shelley's analogy comparing the new system to a smartphone upgrade. She's exactly right. I like to think of myself as a savvy tech user, but it took even me a while to get used to this phone last year," Steve said, waving the phone in his hand.

"Hopefully tomorrow's easier than today was."

"It should be," said Steve. "My team was talking with Brittany, and we're thinking about another training. That should help you from having to explain everything."

"Well, that'd be great!" Jennifer said.

"Thank you for your time, Jennifer," Steve said. "Do you think the boss is still in?"

"You know as well as I do," she said with a wink. "Mr. Mack's always the last to leave."

TVGENIUS

What Does It Really Mean to "Own It"?

Many companies—including those that seem to be at the absolute top of their game—have serious problems that developed somewhere between their business plans and their day-to-day operations and have become lost or hidden in their current systems.

Upgrading those systems will intentionally bring those problems to light and force a review from a fresh perspective.

Now, I understand the enormous implications that may have for some businesses.

But a business with those sorts of internal irregularities is limited in its performance and growth potential—and in this increasingly competitive market, that's a prescription for failure.

You should be aware that designing and implementing an ERP system for your business is a project that requires a significant degree of education, training, and experience.

Whatever your selection process might be, you should exercise the same caution that you would when entering into any other transaction. Ask questions. Ask for references, and check them! Ask more questions.

NEXT STEPS:

Another one of our core values at The Vested Group is "Own It." When you are selecting and designing your new ERP system, it's important that you take as much ownership of this process as the team you hire to guide your implementation. One way we support you in owning your project is through our Define the Why process. Take a look at a sample template at joelpatterson.com/the-book to get an idea of how The Vested Group helps you keep your focus on your purpose and "Own It"!

CHAPTER 16:

MR. MACK AND
THE BIG PICTURE

STEVE WAS CONFIDENT that he'd found his answers—most of them, at least. Before he left the Carter Supply corporate headquarters, he thought he'd stop in and see Mr. Mack.

"You're working late too, huh?" said Dean Mack.

"Good evening, Mr. Mack."

"I like to see that. Success in any business, any endeavor, requires you to work late and put in the time. Too many people have forgotten that these days. They think 'work hours' is something other than a minimum requirement."

Steve nodded diplomatically.

"So, what brings you out to our neck of the woods so late in the day? You find out what's wrong with that system of yours?"

"Thankfully, I haven't found any problems with the system so far."

"I don't like the sound of that, Steve. It's not enough to just put in the late hours—success is based on results. You can't have one without the other."

"What I meant is that we haven't found any problems with the system itself. In speaking with your team, we agreed we should do a bit more training and refine a couple of processes, but we haven't found any actual problems."

"How can that be?" Mack asked. "My people—"

"Need some more training, which is perfectly reasonable, considering the first firm's training didn't apply to our new system. A bit of extra training to unlearn the work of the first firm and relearn this new system."

"Extra training? Sure, but a need for training wouldn't make all my department heads get up in arms."

"Might it be," Steve said gently, "that this change just takes some getting used to?"

Dean Mack paused for a moment, thinking it over. "Come on in and sit for a minute," Mack finally said, offering an invitation into his office with a wide sweep of his arm.

While the hour was already late, something told Steve that this wasn't so much a social offer as it was a request for help.

"My grandfather was quite a man. Used an inventory system he learned from his brother, who managed navy supplies in the Second World War. Could feed ten thousand men or get rations from one Pacific island to the next, day after day, week after week. That system served them and my father well. Me, too, until our competitors started lapping us with the advancements of the digital age and cloud computing. We know it now—the systems of previous generations are no longer enough. There's so much opportunity to be had by embracing business on the cloud. I see that. My team sees that. I guess it's like you said—we live and die by change. It's just taking us a while to embrace the actual change at hand when it comes to how we do our job."

"Well said, Mr. Mack," Steve said. "I understand that all of this new technology can be overwhelming when you're just getting started.

I think the first thing I'd say to you is that you shouldn't beat yourself up because you don't understand it. There's no way that anyone can keep on top of everything as quickly as things are developing these days. But that's the thing: you don't have to."

"I'm listening."

"That's the importance of not *just* having a consultant to come in and do a single job and then get out but rather in having a *partner* that you can rely on over the long haul. You can't look at your computer systems as just one upgrade and then you're done for the foreseeable future. It doesn't work like that anymore. You need to have a partner working with you as you move your business toward what can be an uncertain future the same way that you have a lawyer or an accountant or any other professional. You might not need them every day; you only need them when you need them. So, you don't need to stay at the crest of the technological wave; you just need trusted people who can do that for you. Your grandfather's brother learned his system from the navy at a time when it was state of the art. Now you're learning the most state-of-the-art system from us. We learned it from the best global industries in the world. And these state-of-the-art systems help us all adapt to change, which comes no matter what."

> You need to have a partner working with you as you move your business toward what can be an uncertain future the same way that you have a lawyer or an accountant...

"I see it," Mr. Mack said. "Steve, I'm beginning to understand this partnership idea you've mentioned a few times. I know all too well how change can be almost frightening. But as I always like to say, if you don't like change, you'll like irrelevance even less. My forebears always knew this to be true. If you think about it, there's only one reason my grandfather was able to go from junk collector to successful businessman, and that's because things changed. I built our manufacturing plant because things changed. Change isn't just a part of life; it's a necessary part of life. And often it is where we find the greatest opportunity. Rather than being afraid of change, I see it as a way to maximize opportunities. Which brings me to my point."

Steve nodded in anticipation.

"Steve, you're absolutely right that technology is charging forward at an unprecedented pace. No doubt. And you've got all of the computer companies turning out product that is outdated almost as soon as it's unboxed. There's scheduled obsolescence, right? The tech companies want to keep making new computers, new smartphones, so they make certain that all new hardware only remains current and functioning for a short amount of time. All that waste is piling up. And it's a gold mine. In some cases literally."

"I never thought of it like that."

"Most people haven't. Not yet. But it's an emerging boom market, and we want to be at the forefront of it all. So, anything that happens to cell phones, computers, really anything IT related, we want to use your system to allow a company like ours to process all of that inventory."

"Mr. Mack," said Steve, "I'm excited we're now discussing a way to identify opportunities and then take advantage of them. Why didn't we talk about this before?"

"Before? Getting this system up and running with the company is just a test run for our new approach to inventory management with our new e-waste service line. I had to get to know you and your system better before I showed you the big picture."

"Well, I'm glad you did. I'm in this business because I love finding opportunities and then giving businesses like yours the tools that you'd otherwise be unaware of or overwhelmed by so that you can use those tools to better build and grow your business."

"Well, that will all be possible. As soon as you solve our issues here. And I can tell you, it's not just about change. There's gotta be something more going on than just a few of my people needing more tightly defined processes or extra training. I know my people, and while they're as comfortable with change as the rest of us, they wouldn't let it derail their work. Training is important, sure, but I got a hunch there's something else. Can you solve it for me?"

"Mr. Mack," Steve said, "you have my word."

"And if you solve it for me by tomorrow," Mr. Mack said with a smile, "you have *my word* that we'll be partners on the new business."

TVGENIUS

How Do You Know When You're "All In"?

I think there might have been a day when our services could have been viewed as an in-and-out sort of service. That is, we came in and did this implementation, and then our relationship was over. Times have changed, however. While the software is remarkably self-sustaining and stable, the fact of the matter is that business no longer is. We have clients whose businesses remained virtually unchanged for decades: same suppliers, same products, same vendors, same customers. That used to be regarded as stability and consistency, but today that constitutes stag-nation. We have a current client that warned us when we kicked off the project that they would be growing at a 40 percent annual rate. They grew 50 percent by the time we finished the project five months later. Things move fast, and everyone needs friends that can and, more importantly, genuinely want to help.

And so, our focus isn't limited simply to software implementations. Rather, we're in the business of business improvements. Because of this, we actively develop ongoing relationships with our clients that extend beyond the initial implementation and evolve to a place where we feel like we have phantom ownership in their business. We want to understand where they are headed in the future and how we can help them get there as quickly and efficiently as possible.

NEXT STEPS:

Another one of our core values at The Vested Group focuses on the idea of being "All In." Whether you choose The Vested Group or someone else for your implementation, we'd recommend you look for a partner who embraces the idea of being *all in* and truly walks the walk. Here's a two-minute clip from a client that appreciates our *all in* and then some: joelpatterson.com/the-book.

PART THREE

MYSTERY SOLVED

CHAPTER 17:

THE IMPROBABLE TRUTH

THE NEXT DAY, Steve confirmed that Denise and her team went to Carter Supply to lead additional training for Brittany and her superusers.

He then reviewed the punch list received from the department heads that morning.

And he heard that Jeff and Hannah had reviewed the sandbox.

Steve knew he and his team had done everything they could do. He knew they were remedying every issue that had arisen. But still, something didn't feel right.

Steve shared Mr. Mack's hunch from the night before—the Carter Supply mystery remained unsolved.

Steve knew he had to do something. He knew that every issue had a simple (relatively speaking, that is) explanation, and it was important that they come up with theirs before it got any further out of hand.

And so, without anything more than the determination to arrive at a solution and the confidence that he could do it, he decided to ask Hannah to accompany him for one last ride out to the Carter Supply corporate headquarters.

"Can I ask you for a favor?"

"Is it going to impact the David Martinsson presentation that I'm doing tomorrow?" she asked.

"Oh, that's right—you've got our new client presentation to do tomorrow." Martinsson was as important as any project, but Steve was determined to wrap up the Carter Supply implementation that day. "No, I was hoping you had time to join me for a quick trip out to Carter Supply."

"For sure," she answered.

"I just don't get it," Steve said when they were on their way. "I've been in this business for fifteen years. I've always thought that I was among the best in the whole country, but the answer to this one seems to be escaping me. We've got Denise, Alexa, CJ, and Rian leading training now for Brittany and the superusers. We're updating protocols based on the punch list from the department heads. But there's still the question of why there was a headache at Carter Supply in the first place. I'd be absolutely certain that this was just another example of the Valley of Despair if it weren't for those squirrelly results we've gotten from some of those tests in sandbox. What'd you and Jeff find?"

Hannah shook her head too. "After our talk with Brittany, I caught up with Jeff this morning, and we double-checked the history of what happened in sandbox. It's almost like we weren't making the changes to the system even though we knew that we were making them. That we weren't making the changes, but then we were, and then we weren't again." She shook her head at the difficulty she was having in capturing their situation. "Even within the past forty-eight hours, we saw there was one change: the background color scheme of all things. It was changed, then undone. I can't describe it. Do you know what I mean?"

"I don't ...," he admitted. Then his eyes lit up. "But now I do!"

"You suddenly seem a little more excited and upbeat than usual," Hannah observed. "And that's saying something for you."

"You ever read any Sherlock Holmes?" Steve asked.

"I've seen the movies and TV shows," Hannah replied.

"So you'll know that in his mysteries, Holmes often says, 'When you have eliminated all other possibilities, whatever remains, no matter how improbable, is the truth.' I think that this entire time we've been trying to figure out what the problem is even though everything led us to the inescapable conclusion that there's no problem at all."

"I am still not following what this has to do with your Sherlock Holmes."

"If there's not a problem with the system, but Carter Supply is still experiencing some difficulties, then the only reasonable answer …"

"I'm waiting for it," Hannah said.

Steve dropped his bombshell. "It's user error."

Hannah arched an eyebrow. "So, what now?" she asked.

"I think that we go into Carter Supply and we make the reveal like the greatest detective of all time."

"Like Scooby Doo and Mystery, Incorporated?"

"Them too."

CHAPTER 18:

THE REVEAL

AT THE CARTER SUPPLY CORPORATE CENTER, Steve and Hannah made their way to Mr. Mack's office, where his secretary, Elyse, showed them in to the senior leadership conference room and then followed Mr. Mack's directions to call Ben Mixon, Walker Arnav, Craig Myra, and Brittany Brower to join the assembly.

Steve started. "First, we want to thank you for extending us the opportunity to partner with Carter Supply in the transition over to your new ERP system. It's been our pleasure to serve you, and we sincerely hope that this provides the foundation for an ongoing relationship with TCG."

"Well, before you go planning for anything else," Mr. Mack interrupted, "we better get this current situation under control."

"And that's just exactly why we're here," Steve said. "We've spent the past few days getting to the bottom of the issues you've experienced here at Carter Supply. For the most part, we found what we expected to find. The need for more training because of the difficulty moving from the first firm. The need to add more process definition to areas not in the original list. I'm reminded of Jennifer and her team member Shelley, who said learning this new system is like getting up to speed on a new smartphone. It takes a few weeks to get used to it, but after that, it's so much easier to use than the older model."

Mr. Mack considered the words. "So, you're saying?"

"That the majority of your problems with the system will be resolved if we just *recommit* ourselves to taking the steps that we've already outlined for you in order to get your employees used to working with this new system."

Mr. Mack seemed less than convinced. "You're saying that's all there is to it?"

The others in the room mumbled their own doubts about the proposed solution.

"No, that's not quite all," Steve said. "The one thing that I couldn't understand," Steve continued, "is that when we made attempts to work on the system, we'd come back later and all of that work would be gone."

"So, what does that mean?" Mr. Mack asked. "You just told me that the issue was training and adding process definition, and now you're saying that there's something wrong with the system?"

"No, sir. There's an issue, all right," Steve admitted. "But it's not an issue with the system. It's a human issue."

"Sorry, Steve, but this makes about as much sense as the Longhorns' offensive scheme this year," Craig said.

Walker wasn't in a mood to have his team maligned. "Oh really? You mean it's about as believable as thinking the Sooners have a shot of winning the Big 12?"

"What I'm saying, Walker—and Craig," Steve said, "is that user errors have derailed the system and even the training. And what we discovered doesn't look like an accident."

"Are you suggesting foul play, Steve?" Mr. Mack asked, more alarmed than shocked—but still plenty shocked. "Our security is top notch. There's no one getting in here without our knowing about it."

"That's exactly right," Steve said. "I don't think the user error is caused by an outsider. I think it's someone within Carter Supply. In fact, someone right in this very room."

"Who would do such a thing?" Mr. Mack wondered aloud. "And why?"

"Allow me to explain," Steve said. "When we develop an ERP system, we actually develop two systems. One is the actual system that is going to do the day-to-day work, which we call production. The other is a duplicate system that we like to call sandbox. It's like a demo or test system, if you will. This way, we can make changes to the test system, sandbox, to determine how they will interact with the system and refine them before they are actually incorporated into the production environment. It's like making sketches before you actually sit down to paint a canvas."

"That's clear," said Mr. Mack.

"In order to distinguish the two systems from one another so that we don't get them mixed up, we color code them, because you can imagine the disastrous confusion that would result if sandbox and the production 'live' system that our client is going to utilize were to get mixed up with one another."

Craig looked at Walker but said nothing.

Walker looked at Craig but said nothing.

"If you remember," Steve continued, "when we started the project, there was quite a controversy as to what color we were going to choose to represent the real 'live' system and which color would stand for sandbox."

"There wasn't a controversy," Walker insisted. "We live in Texas. The color for the real system should have been Longhorn burnt orange."

"Not if we wanted a winner," Craig insisted. "It should have been Sooner crimson red. Best color. Best team."

Steve interjected, "When our team discussed the project this week, we realized something was amiss. Whenever we had made a change to sandbox, it would either disappear or, worse yet, would have some impact on the actual system. We couldn't figure it out. But now I understand that what was happening had nothing to do with the system itself. It was the *color* representing production—and the color representing sandbox."

Craig and Walker flashed looks at one another again.

Craig was the first to speak. "I guess I didn't understand the importance of the two systems. I admit that I was really mad when you capitulated to Walker and decided to make the designation for the live system Longhorn burnt orange. Everyone knows that Sooner crimson is the color of winners. So, I came into the office one night and changed the color from orange to crimson."

"You did what?" Mr. Mack asked.

"The joke's on you," Walker said. "Because I knew how steamed you were that they chose to go with the better color—burnt orange—I figured that you'd do something like that, so the first thing I did that moment was to check the system and change them back. So you see, I didn't do anything wrong—I corrected the problem."

"You did what?" Craig asked. "Because I felt guilty about what I'd done, I went and changed them back."

"You did what?" Walker demanded.

"I thought someone might find out what I had done and think I was tampering with it. Besides, I thought it would be better if you got busted for it. So, I went and changed it back."

"But I was worried that I had left some evidence," Craig said. "So, I went back and turned it back again."

"And that's your explanation," Steve said, turning to Mr. Mack. "The identifying colors for the live system and our sandbox kept going back and forth. Sometimes people were working in the live system, but at other times they were just wasting time in sandbox. Back and forth. And that's the cause of it all: the age-old feud between Longhorns and Sooners."

"You mean Sooners and Longhorns," Craig corrected.

"Hell, I'm a Razorback through and through," said Mr. Mack, "but you're telling me that all of the problems that we've been having are because of these two going back and forth on some kind of frat-house prank?"

Steve shrugged and nodded. "That appears to be the case."

"No," Ben Mixon said, stepping forward. "It was me, too. I turned the color coding to maroon for Texas A&M. I admit it and I'm not ashamed. Go Aggies!"

Brittany Brower shifted uneasily and then leaned over next to Ben. "Oklahoma State. Cowboy orange," she stated proudly.

"Well, I'll be," Mr. Mack said. "If that's not the craziest thing that I ever heard. All of this turmoil to this business, and it was just a couple of jokers having some fun and supporting their football team."

Everyone looked to the old man to see what his reaction would be.

Mack just laughed. "I don't suppose most men—or women— could find anything wrong with that."

Craig and Walker laughed, too.

"Are you sure, Mr. Mack?" Walker asked.

"You're not mad?" Craig followed.

Mr. Mack kept laughing. "I said *most* couldn't find fault with you two. But like I said, I'm a Razorback through and through, and I can't believe what your foolishness has done to the company. I should fire you all. Go clean up your mess."

The four exited the room while Steve and Hannah stayed behind.

"So that's it, Mr. Mack," Steve said. "I certainly wish it won't come to anyone getting fired."

"Oh, they'll be fine," Mr. Mack said. "They aren't fired. But there's no reason why they shouldn't stew in their own juices for a little bit."

> There is a lot to be gained in consulting with us over the course of time as you grow your business, because your needs for this system are going to change accordingly.

"Well then, I guess that's a happy ending after all," Steve said, smiling at Hannah. "The system is in place just like we designed, and I think if your people just give themselves some time to get acclimated and follow the training procedures that we left for them, you'll soon find that your business will realize all of the benefits that we promised would be delivered by the system."

"I certainly hope so."

"And as I said, this doesn't need to be the end of our relationship," Steve said. "There is a lot to be gained in consulting with us over the course of time as you grow your business, because your needs for this system are going to change accordingly. We can help you leverage the system to evolve along with your business, and we can also assist you in ways that the system can actually be the impetus for business growth."

"I'll be in touch with you, then," Mr. Mack said. "And till then, I'm much obliged."

"It's been our pleasure," Hannah said.

Steve and Hannah walked down the hall.

"It all worked out all right," Steve said. "And it's a job that we certainly won't forget for a long, long time."

"Agreed. Every mystery should end that way."

TVGENIUS

There.

Just as promised. A comedic mystery with a happy ending.

And I hope that you learned a little something about The Vested Group and how we can help your business maximize today's technology in order to reach your greatest potential and achieve all of your goals.

But don't stop turning pages just yet.

There's still more to come ...

CHAPTER 19:

THE BIG COMMITMENT

"YOU'RE SURE YOU want me to handle this?" Hannah asked.

"You're ready." Steve nodded and smiled. "And if you find that you need help …"

"Got it. Thanks," Hannah said, stepping through the open conference room door.

"Mr. Martinsson, so good to see you," she said as she offered an outstretched hand to TCG's newest client. "First, we want to thank you for the opportunity to represent the Green Legacy. We've already reviewed the documents that have been provided to us, and our expert consultants have compiled a complete breakdown of your business and designed a system to accommodate your strategic objectives while positioning your company to scale into the future."

"That sounds fine," said Mr. Martinsson.

"Our next step will be what we call our Phase Zero, which is a process unique to TCG. During your Phase Zero, we will walk you through all of the planning and analysis steps to make certain that you know exactly what's going to take place and to ensure your company has a successful implementation."

"That's important to us," Mr. Martinsson said.

"Once we begin the actual implementation, you can rest assured that with our experience and project management skills, we'll complete a successful project that is tailored to the specific needs of your business.

> At TCG view every ERP implementation not as a project but as a partnership, with your company and ours working together closely to accomplish mutually shared goals.

In fact, we just recently wrapped up a project that was one of those once-in-a-career experiences—I hope—and a real demonstration of what TCG can do for their clients." She chuckled to herself. "But that's another story for another time."

"I look forward to hearing it," Mr. Martinsson said.

"It is kinda hysterical," she said before moving on. "But the important thing we all realized is that we at TCG view every ERP implementation not as a project but as a partnership, with your company and ours working together closely to accomplish mutually shared goals."

"That's refreshing," Mr. Martinsson said.

"It is," Hannah said. "Because we depend on deep trust. We know that the implementation is going to cause some level of disruption in your current business, but our goal is to minimize those impacts. And we need you to trust us on that, keep communications open, and never, ever wander off unknowingly into the Valley of Despair."

"The Valley of Despair?" Mr. Martinsson asked.

"Don't worry, you'll get there," Hannah said and then with a confident smile added, "And when you do, we'll walk you through it and guide you out."

"So you're saying," Mr. Martinsson said, "that ERP implementations aren't easy. In fact, they're complicated and require a lot of change. But you'll help us every step of the way with whatever we need, because we're partners in this?"

"Exactly. All it takes from you is a big commitment. Can you do that?"

"Certainly. Just one question before we dive in."

"Shoot."

"Just what does TCG stand for?"

"'*Tell* the *Competition. Goodbye.*'"

TVGENIUS

Reduce, Reuse, Recycle ... *reVESTED?*

While the company in this chapter, Green Legacy, is fictional, our business management solution for companies in the e-waste and reverse logistics space is quite real. We weren't kidding when we said we are committed to the ideas embodied by Conscious Capitalism (www.consciouscapital-ism.org)—so much so that we've spent several years developing a solution that caters to companies that process electronic waste and/or fall into the reverse logistics category. Our solution is called reVESTED—the re comes from the popular Reduce Reuse Recycle slogan, and the VESTED comes, of course, from The Vested Group. You can find out more about reVESTED and how it works at www.thevested.com/revested.

NEXT STEPS:

If you are curious about why we are so proud of our reVESTED solution and why we believe supporting the e-waste and reverse logistics industry is so important, take a look at our documentary on the topic, Silicon Mountain at joelpatterson.com/the-book.

CHAPTER 20:

NEW BEGINNINGS

AND THEN THERE came a day that Steve had thought might never come, the culmination of all of his and his team's efforts over the years: the official opening of TCG's new offices.

There was still a skeleton crew of construction workers on the site putting all of the little finishing touches on the place that morning.

With another general contractor, Steve might have been impatient with their presence, but he was glad to have his dad at the property for one last day. Gladder still to see his mother had added the details to the interior decorating. Sconces on the walls, blinds on the windows, a color scheme that perfectly matched what he'd envisioned.

Steve made a beeline to his parents, who stood together looking at the finished product. "You really did a great job, Mom and Dad."

"Of all the jobs I've ever done in my career—and I've done a whole helluva lot of them," said his father, "I think this project is the one I'm most proud of. Someday—a long, long time from now, I hope—when my buddies are remembering me and talking about my work, I hope this building is the one they point out when they talk about the work I did."

"What we are proudest of," said his mother, "is that we got the opportunity to help build a house for the wonderful dream that you've brought to life with your company. I won't ever get the chance to say it enough, but I'm enormously proud of you and all that you've built—brick by brick, just like we taught you. And I am so grateful that I got the opportunity to create a building to house that dream, to give you some place to grow it even bigger. And bigger."

They didn't say anything more than that.

Although Steve may have added a soft "Thanks" almost under his breath.

And that was enough.

In a chain of events that was somewhat unusual for the folks at TCG, the old bar that they had decided to leave in their offices as a workstation was converted for the night into, well, a bar. Pitchers of margaritas and daiquiris were refilled almost as soon as they were emptied—which was very quickly.

And, of course, there was a metal tub of Shiner Bock longnecks on ice, because it was Texas, after all.

Most important of all, however, there was a very large crowd. The guys from the construction crew came back for the event.

Steve and Jeff. And Maria, of course.

All of the teammates.

There was a wonderful showing from the clients that they had served over the years.

Friends and family.

It was everything that Steve had always wanted it to be.

And then, just as Steve didn't think the party could hum any better, he was thrilled to see Mr. Dean Mack walk in, with Brittany, Craig, Walker, and Ben in tow. Steve ran over to them.

"Mr. Mack," Steve said after shaking hands with the group. "I can't thank you enough for coming to the party."

"And I can't thank you enough for solving our mystery. Considering the circumstances," Mr. Mack said, gesturing at his group, "I thought we all needed to take a break and enjoy ourselves."

"We have plenty to celebrate," said Steve.

"Sure we do," said Mr. Mack. "We're partners, aren't we? There for the bad, and there for the good. So let's enjoy the good, why don't we?"

Hannah stepped in to lead Craig to a group of TCG Sooners supporters while Steve helped Walker find the company's Longhorns cohort.

When it finally got to that point in the evening when drinking glasses began to be clinked with silverware, it was Jeff who was the first to move to the head of the crowd and offer some words to mark the occasion.

"Seriously, I just want the opportunity to thank you all for being a part of this amazing adventure. And that's what it's been. It started out as a business proposal for a consulting group that would primarily oversee the implementation of cloud-based software for corporate clients. What could be a bigger bore than that?"

The room laughed.

"But I'm telling you that every single day, I get up and I'm excited to come to work, because I know that's exactly what it's going to be: an adventure.

"A crazy, occasionally maddening, always thrilling adventure. And I have all of you to thank for that. Because it's not the amazing opportunities that our valued clients give us that necessarily make that come true for me; it's all of you. Our teammates. It's getting to share this." He gestured around the room. "All of this. With all of you. That's what makes this job the adventure that drives me every day. And I need to thank you for that. Most of all, I want to thank Steve for bringing us together."

There was a big round of applause as the room clapped for Jeff, the team, and Steve. And then, of course, it was Steve's turn to speak.

He had spent a good deal of time preparing remarks for the occasion, but now that the time had come, he wasn't sure at all that he had anything to say. Not really.

Maria knew better.

She kissed him, smiled, and pointed him where he needed to be.

Steve walked to the head of the crowd, and a hush fell over the room.

A momentary wave of emotion swept over Steve, and he thought for the briefest of moments that he might not be able to say anything at all. But from the back of the room, his father and mother nodded at him, and suddenly, with that simple gesture, he knew exactly what he needed to say.

"Thank you. For everything."

To those standing near Steve, his eyes seemed to suddenly mist, as if he was overcome with emotion. But they simply misunderstood that he'd just gotten some leftover construction dust or debris momentarily lodged in his eye that needed to be wiped away. That was it—just construction dust in the air.

"To all of the clients who were gracious enough to join us here tonight, thank you. For your trust and your commitment—"

"And our checks?" one of them called out good-naturedly.

Steve laughed with the others. "Yes, you've all made our get-together tonight possible in so many ways."

More laughter.

"But seriously, we are—all of us at TCG—aware of how overwhelming our services can be for our clients and their businesses. It's a necessary part of the process, but it works. We know that each and every one of you have trusted us with the businesses that you've tended to and grown—your life's work. And so, while it was a pleasure doing business with you all, we want you to know—we hope you always knew—that it was more than just business to us.

"And that's how we went about our business—with an absolute commitment that it was so much more than just that."

There was a polite round of applause.

"From day one, that was our idea: a consulting group that would offer the very best in technical expertise and services but look past the details to see and consider the human element, as well. Maybe more so. What we set out to accomplish from the very beginning was to provide professional services that were every bit as good as those that anyone else in the industry could offer but coupled with a *commitment* to our people and our client's business that was far, far better.

"I believe we've accomplished that, and I want to thank all of you who partnered with us and allowed us that opportunity. We treasure the relationships that we've formed, and we look forward to partnering with you in helping you grow your business and exceed your expectations for many, many years to come."

All the team members offered an appreciative round of applause.

"And that's not all that I'm proud of. In the same way that we recognized from the very beginning that we were asking our clients to trust us with something special—their life's work—we do the same with each and every one of you. You're not just our employees, our team members. You're our representatives. Every one of you is the face of this company ... and I can't find the words right now to tell you all just how proud that makes me.

"Every single one of you has demonstrated time and time again that this is more than just a job, more than just a paycheck. You have all been so incredibly generous with yourselves, with your time and talents, and with your warmth and friendship."

There was more applause. Steve searched out his father again, and they exchanged nods this time.

"Lastly—my father is a builder by trade. And my mother an interior decorator. Together, they taught me well the difference between a house and a home. I look around at all of you, and I see something more than employees, more than coworkers, something more even than friends. You've all come into a workplace and made it into our home."

And some more of that "construction dust" may (or may not) have found its way, again, into the eyes of the many assembled.

"Thank you," Steve said.

A number of those in the crowd called back their own "Thank you."

"Friends and family. Clients and community. Partners and team members. I owe so much to so many," Steve said. "So please join me in raising a glass to TCG. The Caring-est Group of people you'll ever have the pleasure of working with."

* * *

When the hour had grown late, all the food had been eaten, and most of the drinks had been drunk, the crowd began the thin out. Appropriately enough, their departure was like the development of TCG, but in reverse.

The friends and clients all said their goodbyes, expressed their thanks, and then said good night.

The newest hires were the next to leave—one by one—until all that remained were the handful that had been there from the very beginning.

Maria came up from behind Steve and slid her arm around his waist. "It's a pretty ragtag group," she said.

"It was, at that."

She smiled. "And you're the worst of them all. I can't imagine what I was thinking. Any chance for a do-over?"

He put his arm around her shoulder. "Not a chance in the world."

"Then what's wrong?" she asked him.

"Do you think it was a mistake? All of this? Growing the company?" Steve asked.

"We've been over this," she reminded him.

"I know, but I treasure these people. What we've built ... well, it is the answer to that dream I had. We made that real, and I get to live it every day. That's a rare, rare thing, and I don't want to screw it up. Adding more people to the team—"

"Lets us help more clients," she finished for him. "That's a good thing."

"I suppose so."

"And it gives us more teammates. More of this. That's also a good thing."

"I just don't want the special thing that we've built with TCG to get diluted."

"Diluted?"

"You know—sometimes the bigger a company gets, the more they lose touch with those core values. It's really important to me that not only do we continue to serve our clients in that way but that we still work with one another like that, too. I don't want to lose what we've built."

"And we won't," she promised. "The core values aren't just sayings to us, and that's not going to change. Not with fifty consultants. Or a hundred. Or—"

"A hundred?"

"Relax. This bigger building is just a new space for the same TCG. This isn't the end of an era; it's the beginning of a bright new future. That's the point of what TCG really stands for, right?"

"What's that?"

"Not just The Caring-est Group, or the Tech Consultant Gurus. But whatever beautiful thing we want it to."

TVGENIUS

Can You Identify Where You Find Joy?

We'll wrap up this story with a bit about our final core value: "Enjoy the Ride." The Vested Group doesn't tout the tired "Work Hard, Play Hard" slogan. We prefer to lean on strategy and experience when it comes to work. When it comes to play, we don't think there should be anything hard in that. And while we don't like to brand ourselves simply as "fun" either, we do find what we do and the road we take to our shared successes—with team members and clients alike—to be quite enjoyable. So we've found that "Enjoy the Ride" suits us—it speaks to the daily delights, big and small, we look for and find in the journey we are taking together and hopefully someday will take with you as well!

NEXT STEPS:

Find joy in your next "big commitment" by partnering with us! Contact The Vested Group today!

ABOUT THE VESTED GROUP

THE VESTED GROUP is your NetSuite Partner and offers a full range of ERP (Enterprise Resource Planning), EBC (Enterprise Business Capabilities), CRM (Customer Relationship Management), and eCommerce services for growing companies.

We work with a variety of industries including Manufacturing, Wholesale Distribution, Food and Beverage, Professional Services, Grain Milling, Oilfield Services, and Electronic Waste Recyclers and Redistributors.

The Vested Group provides services in the following areas:

- NetSuite Licensing
- NetSuite Pre-Implementation Planning—Phase Zero
- NetSuite Free Trial
- NetSuite Implementation
- NetSuite Support
- Built for NetSuite Certified Solutions
- NetSuite Customization

- NetSuite Development
- reVESTED—the business management solution for the electronics recycling industry, built on the NetSuite platform by The Vested Group

The Vested Group is a Five-Star NetSuite Partner with a team of NetSuite Certified Technical and Functional Consultants based in the Dallas, TX area. Additionally, The Vested Group is a Certified NetSuite Commerce Partner and Certified SuiteSuccess Partner and founded the Dallas Fort Worth NetSuite User Group.

The Vested Group has been featured on the Inc. 5000 America's Fastest-Growing Private Companies List in 2017, 2018, and 2019 as well as the Inc. Magazine Best Workplaces in 2018, 2019, and 2020. The Vested Group was awarded second place in the Dallas Business Journal Best Places to Work in 2018 and 2019.

Please visit www.thevested.com to learn more about our amazing organization.